MASTERING ESSENTIAL MATH SKILLS

No-Nonsense Algebra
Master Algebra the Easy Way!

Richard W. Fisher

*America's Math Teacher, Richard W. Fisher,
will carefully guide you through each and every topic
with his award-winning system of teaching.*

Go to **www.NoNonsenseAlgebra.com**
for instant access to the Online Video Lessons.

Your access code is: **D2GH7Y4M3**

Math
Essentials
LOS GATOS, CALIFORNIA

Dedicated to my lovely wife, Elena,
and to my beautiful daughter, Vica.
Thank you for the inspiration.

Mastering Essential Math Skills: No-Nonsense Algebra, 2nd Edition

Copyright © 2018, Richard W. Fisher. Printed and bound in the United States of America.

For information, please contact Math Essentials, P.O. Box 1723, Los Gatos, CA 95031.

Although the author and publisher have made every effort to ensure the accuracy and completeness of information contained in this book, we assume no responsibility for errors, inaccuracies, omissions, or any inconsistency herein. Any slighting of people, places, or organizations is unintentional.

First printing 2018

ISBN 978-0-9994433-3-0

Introduction

ABOUT THIS BOOK

What sets this book apart from other books is its approach. It is not just an algebra text, but a **system** of teaching algebra. Each of the short, concise, self-contained lessons contain five key parts:

1. A clear **introduction and explanation** of each new topic, written in a way that is easy for the student to understand
2. A **Helpful Hints** section that offers important tips and shortcuts
3. **Examples** with step-by-step solutions
4. **Written Exercises** with answers in a the back of the book
5. A **Review** section that ensures that the student remembers what has been learned

Each lesson contains the necessary structure and guidance that will ensure that the student will learn algebra in a systematic, step-by-step, logical manner. The book is set up in chapters, and there is a natural flow from each lesson to the next. At the end of each chapter there is a **Chapter Review** section. A comprehensive **Final Exam** is also included.

HOW TO USE THIS BOOK

Step 1

Carefully read the **Introduction** that begins each lesson. This section will include essential information about each new topic. Here you will find explanations as well as important terms and definitions.

Step 2

Carefully read the **Helpful Hints** section. This section will provide important tips and shortcuts. When completing the written exercises, it is often helpful to refer back to this section.

Step 3

Carefully go through the **Examples**. Each example shows the step-by-step process needed to complete each problem

THE MOST IMPORTANT TIP THAT I CAN OFFER WHEN USING THIS BOOK!

Copy each example on a piece of paper. Than read and copy each of the steps. I promise that by doing this, you will find it much easier to understand the problem.

There is something very special about writing a problem down and then writing out the steps. It makes the learning process so much more effective. When you do this, you are fully involved and will have a much deeper understanding. Just simply reading a problem and the steps is not nearly as effective.

Step 4

Work the written **Exercises**. If necessary, go back and re-read the **Introduction** and the **Helpful Hints**. You may want to go back and refer to the **Examples**, also.

Step 5

Complete the **Review** section. The review problems will ensure that you remember what you have learned.

Step 6

After completing the **Exercises** and **Review**, correct your work. The **Solutions** section is located in the back of the book.

HOW TO USE THE FREE ONLINE VIDEO LESSONS

www.nononsensealgebra.com

Go to **www.nononsensealgebra.com** and you will find a corresponding video lesson for each lesson in the book. The author, award-winning teacher, Richard W. Fisher, will carefully guide you through each topic, step-by-step. Each lesson will provide easy-to-understand instruction, and the student can work examples right along with Mr. Fisher. It's like having your own personal math tutor available 24/7. After the video lesson, you can go to the book and complete the lesson. The book combined with the video lessons will turbo-charge your ability to master algebra.

Your access code
to the online lessons is located
on page 1 of this book.

Table of Contents

Chapter 1: Necessary Tools for Algebra

Chapter 2: Solving Equations

Chapter 3: Graphing and Analyzing Linear Equations

Chapter 4: Solving and Graphing Inequalities

Chapter 5: Systems of Linear Equations and Inequalities

Chapter 6: Polynomials

Chapter 7: Rational Expressions (Algebraic Fractions)

Chapter 8: Radical Expressions and Geometry

Chapter 9: Quadratic Equations

Chapter 10: Algebra Word Problems

Mastering Essential Math Skills: Book 1/Grades 4-5

Mastering Essential Math Skills: Book 2/Middle Grades/High School

Whole Numbers and Integers

Fractions

Decimals and Percents

Geometry

Problem Solving

Pre-Algebra Concepts

No-Nonsense Algebra Practice Workbook

Try our free iphone app, Math Expert from Math Essentials

For more information go to www.mathessentials.net

Some Final Tips Before You Get Started

- Before you start each lesson, always have a few sharp pencils, a ruler, as well as regular lined paper and graph paper.

- Start each lesson by watching the Online Tutorial Video that corresponds with the lesson in the book. Don't be passive. Work right along with the instructor, showing all your work. You will learn a lot more this way.

- Next, carefully read the Introduction that begins each lesson.

- After that, carefully read the Helpful Hints section.

- Next come the Examples. For maximum benefit, don't just read the examples, but copy each example and all of the steps. There is something very special about writing down a problem and all the steps. It helps you to learn much more effectively.

- You are now ready to work the Exercises. Make an effort to write as neatly as possible, and show all of your steps. If you experience difficulties, remember to go back to the Introduction or the Helpful Hints section for help. Also, it might be good to go through the Online Tutorial Video a second time.

- The last part of each lesson is the Review section. Learning requires repetition. Just ask any athlete or musician. You have to practice to get good at anything. This section will provide the practice necessary to remember what you have learned.

- Once you have completed the entire lesson, use the Solutions section in the back of the book to correct your work. It is important to re-work any problems that were worked incorrectly. You need to find out what caused your mistakes. That way, you can avoid making the same mistake in the future.

- After following the tips on this page, you are ready to move on to the next lesson.

1-1 Adding Integers

INTRODUCTION

Integers are the set of whole numbers and their opposites.

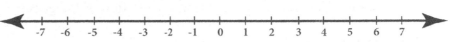

Integers to the left of zero are negative and less than zero. Integers to the right of zero are positive and greater than zero. When two integers are on a number line, the one farthest to the right is greater. Remember the following when adding integers.

Helpful Hints
- When adding integers, always find the sign of the answer first. Any number with no sign is assumed to be positive.
- The sum of two negatives is a negative.
- When adding a negative and a positive, the sign of the answer is the same as the integer farthest from zero. To get the answer, subtract.
- When adding more than two integers, group the negatives and positives separately, then add.

EXAMPLES

1) -7 + -5 = -
    (the sign is negative)

$$\begin{array}{r} 7 \\ +\ 5 \\ \hline -12 \end{array}$$

2) -7 + 9 = +
    (the sign is positive)

$$\begin{array}{r} 9 \\ -\ 7 \\ \hline +2 \end{array}$$

3) -6 + 4 + -5 =
    Combine the negatives
    -11 + 4 = -
    (the sign is negative)

$$\begin{array}{r} 11 \\ -\ 4 \\ \hline -7 \end{array}$$

4) 15 + -8 + 12 + -6 =
    Combine the negatives and positives separately
    -14 + 27 = +
    (the sign is positive)

$$\begin{array}{r} 27 \\ -\ 14 \\ \hline +13 \end{array}$$

EXERCISES

Add each of the following:

1) -15 + 29

2) -12 + -6

3) 42 + -56

4) -39 + 76

5) -96 + -72

6) -86 + 73

7) -9 + 6 + -4 + 3

8) -16 + 32 + -18

9) -32 + 16 + -17 + 8

10) -89 + 27 + -76

11) -16 + -18 + 72 + -12

12) -27 + -19 + -32

13) -329 + 219

14) -617 + 296

15) -509 + -347

1-2 Subtracting Integers

INTRODUCTION

Subtraction of integers is quite similar to addition of integers.

Helpful Hints
- To subtract an integer means to add its opposite. For example, 7 – 8 means the same as 7 + -8.
- Don't let double negative signs confuse you. For example, 9 – -12 means the same as 9 + 12.
- Think of subtraction as addition in disguise.

EXAMPLES

1) $15 - 27 =$
   $15 + -27 =$ (sign is negative)
   $$\begin{array}{r} 27 \\ - 15 \\ \hline -12 \end{array}$$

2) $-3 - -8 =$
   $-3 + 8 = +$ (sign is positive)
   $$\begin{array}{r} 8 \\ - 3 \\ \hline +5 \end{array}$$

3) $-16 - 19 =$
   $-16 + -19 =$ (sign is negative)
   $$\begin{array}{r} 16 \\ + 19 \\ \hline -35 \end{array}$$

EXERCISES

Subtract each of the following.

1) -6 – 8

2) 6 – 9

3) 15 – 18

4) 3 – -9

5) -16 – -25

6) -16 – 12

7) 32 – -14

8) -35 – 14

9) -6 – 4

10) -64 – -53

11) -49 – 54

12) -63 – -78

13) -12 – 16

14) -9 – -18

15) 5 – -25

16) 19 – 76

17) -25 – -76

18) -79 – 147

REVIEW

Simplify each of the following:

1) -24 + 12

2) 15 + -8

3) -13 + -36

4) -9 + 13 – 19 + 4

1-3 Multiplying Integers

INTRODUCTION

Multiplying integers is similar to multiplying whole numbers. Remember the following.

Helpful Hints
- The product of two integers with different signs is negative.
- The product of two integers with the same signs is positive.
- When multiplying more than two integers, group them into pairs to help simplify.
- There are different ways to show multiplication. Look at these examples.

$$5 \times 8 \; = \; 5 \cdot 8 \; = \; 5(8) \; = \; (5)(8)$$

EXAMPLES

1) $7 \cdot -16 = -$ (sign is negative)

$$\begin{array}{r} 16 \\ \times\, 7 \\ \hline -112 \end{array}$$

2) $-8(-7) = +$ (sign is positive)

$$\begin{array}{r} 8 \\ \times\, 7 \\ \hline +56 \end{array}$$

3) $2 \cdot -3(-6) =$
$(2 \cdot -3)\,(-6) =$
$(-6)\,(-6) = +$ (sign is positive)

$$\begin{array}{r} 6 \\ \times\, 6 \\ \hline 36 \end{array}$$

4) $-3 \cdot -4 \cdot 4 \cdot -5 =$
$(-3 \cdot -4)\,(4 \cdot -5) =$
$12 \cdot -20 = -$ (sign is negative)

$$\begin{array}{r} 20 \\ \times\, 12 \\ \hline -240 \end{array}$$

EXERCISES

Multiply each of the following.

1) -3×16
2) $-18 \cdot 7$
3) $-4 \cdot -17$
4) 16×-4
5) $-24(-12)$
6) $(23)(-16)$
7) -23×32
8) $(-2)(-3)(-4)$
9) $-8(-2) \cdot 3(-4)$
10) $6(-4) \cdot 3(-4)$
11) $(-3)(-2)(3)(4)$
12) $10(-11)(-3)$

REVIEW

Simplify each of the following.

1) $-20 + -12$
2) $63 + -27$
3) $7 - 16$
4) $-13 - 15$

1-4 Dividing Integers

INTRODUCTION

The rules for dividing integers are the same as the rules for multiplying them. Remember that the word **quotient** means division.

Helpful Hints
- The quotient of two integers with different signs is negative.
- The quotient of two integers with the same signs is positive.
- When finding the quotient, determine the sign of the answer, and then divide.

EXAMPLES

1) $36 \div -4 = -$ (sign is negative) $4 \overline{)36}$ → -9

2) $\dfrac{-123}{-3} = +$ (sign is positive) $3 \overline{)123}$ → $+41$

3) $\dfrac{-36 \div -9}{4 \div -2} = \dfrac{4}{-2} = -2$ (sign is negative)

4) $\dfrac{4 \times -8}{-8 \div 2} = \dfrac{-32}{-4} = +8$ (sign is positive)

EXERCISES

Divide each of the following.

1) $-36 \div 9$

2) $\dfrac{-90}{-15}$

3) $-64 \div 4$

4) $-336 \div -7$

5) $\dfrac{-75}{-5}$

6) $104 \div -4$

7) $\dfrac{54 \div -9}{-18 \div -9}$

8) $\dfrac{16 \div -2}{-1 \times -4}$

9) $\dfrac{-75 \div -25}{-3 \div -1}$

10) $\dfrac{42 \div -2}{-3 \bullet -7}$

11) $\dfrac{45 \div -5}{-9 \div 3}$

12) $\dfrac{-56 \div -7}{-36 \div -9}$

REVIEW

Simplify each of the following.

1) $-3 - -6$

2) $-5 \bullet -6$

3) $(-3)(7)(-2)$

4) $-66 + -37$

1-5 Positive and Negative Fractions

INTRODUCTION

The rules for positive and negative fractions are the same rules that are used for integers. Remember the following when working with positive and negative fractions.

Helpful Hints
- Determine the sign of the answer before completing the work.
- The sum of two negative fractions is negative.
- When adding a negative and a positive, the sign of the answer is the same as the fraction farthest from zero. To get the answer, subtract.
- To subtract a fraction, just means to add its opposite.
- To compare the values of any fractions, it is good to find a common denominator.
- When multiplying or dividing two fractions with different signs, the answer is negative.
- When multiplying or dividing two fractions with the same signs, the answer is positive.

EXAMPLES

1) $-\frac{1}{2}+\frac{3}{5}$ *Find the least common denominator.*

$-\frac{5}{10}+\frac{6}{10}=+$ (sign is positive)

$\frac{6}{10}-\frac{5}{10}=\boxed{\frac{1}{10}}$

2) $\frac{1}{3}-\frac{1}{2}$ *Find the least common denominator.*

$\frac{2}{6}-\frac{3}{6}=-$ (sign is negative)

$\frac{3}{6}-\frac{2}{6}=\boxed{-\frac{1}{6}}$

3) $-\frac{3}{5}\times 1\frac{1}{3}=-$ (sign is negative)

$\frac{3}{5}\times\frac{4}{3}=\boxed{-\frac{4}{5}}$

4) $-\frac{2}{3}\div-\frac{1}{2}=+$ (sign is positive)

$\frac{2}{3}\times\frac{2}{1}=\boxed{1\frac{1}{3}}$

EXERCISES

Simplify each of the following.

1) $-\frac{1}{5}+\frac{1}{2}$

2) $\frac{1}{2}+-\frac{2}{5}$

3) $\frac{1}{2}-\frac{3}{4}$

4) $-\frac{2}{3}+-\frac{1}{2}$

5) $-\frac{4}{5}\times 2\frac{1}{2}$

6) $\frac{5}{8}+-\frac{1}{4}$

7) $-\frac{1}{3}-\frac{1}{4}$

8) $-1\frac{2}{3}\times-1\frac{1}{2}$

9) $-\frac{3}{4}\div\frac{1}{3}$

10) $\frac{5}{8}+-\frac{1}{2}$

11) $-2\frac{1}{4}\div-\frac{1}{4}$

12) $-\frac{1}{5}+\frac{2}{3}$

REVIEW

Simplify each of the following.

1) 7×-9

2) $16--9+2$

3) $-48\div-2$

4) $\dfrac{-20\div 2}{10\div-2}$

1-6 Positive and Negative Decimals

INTRODUCTION

The rules for positive and negative decimals are the same rules that are used for integers. Remember the following when working with positive and negative decimals.

Helpful Hints
- Determine the sign of the answer before completing the work.
- The sum of two negative decimals is negative.
- When adding a negative and a positive, the sign of the answer is the same as the decimal farthest from zero. To get the answer, subtract.
- To subtract a decimal, just means to add its opposite.
- When multiplying or dividing decimals with different signs, the answer is negative.
- When multiplying or dividing two decimals with the same signs, the answer is positive.

EXAMPLES

1) -.71 + .9 = +   .90
   (sign is positive) − .71
   ──────
   (.19)

2) -2.9 − 3.2 = -   2.9
   = -2.9 + -3.2 (sign is negative) + 3.2
   ──────
   (-6.1)

3) -.5 x 1.23 = -   1.23
   (sign is negative) x .5
   ──────
   (-.615)

4) -3.12 ÷ -.3 = +
   (sign is positive)   (10.4)
   .3 ⟌ 3.12

EXERCISES

Simplify each of the following.

1) -3.21 + 2.3

2) 5.15 ÷ -.5

3) -5.2 − 7.61

4) 5.63 + -2.46

5) -.7 x 6.12

6) 5.9 − -6.23

7) -7.11 ÷ -3

8) -.72 + .9

9) -2.13 x -.2

10) 6.2 + -.73

11) 5.2 + -3.19

12) -5.112 ÷ .3

REVIEW

Simplify each of the following.

1) $\frac{2}{3} + -\frac{1}{2}$

2) $\frac{1}{3} - \frac{2}{3}$

3) $-\frac{2}{3} + -\frac{1}{2}$

4) $\frac{3}{4} \times -1\frac{1}{2}$

1-7 Exponents

INTRODUCTION

An **exponent** is a number that indicates the number of times a given **base** is used as a factor. In the expression n^2, n is the **base**, and 2 is the **exponent**. In the expression 5^3, the number 5 is called the **base** and the number 3 is called the **power** or **exponent**. The exponent tells how many times the base 5 is to be multiplied by itself. In other words, you would multiply 5 by itself three times: $5^3 = 5$ x 5 x $5 = 125$.

Negative numbers can have exponents: $(-2)^3 = (-2)$ x (-2) x $(-2) = -8$.

Remember the following when working with exponents.

Helpful • Any number to the power of one equals that number. For example, $n^1 = n$,
Hints and $5^1 = 5$.

• Any number to the power of zero is equal to one. For example, $n^0 = 1$, and $5^0 = 1$.

• Any number without an exponent is assumed to have one as its exponent. For example $8 = 8^1$.

• Many conventional numbers can be written as exponents. For example $25 = 5^2$, and $16 = 4^2 = 2^4$.

EXAMPLES

1) $6^2 = 6$ x $6 = \boxed{36}$ 2) $(-3)^3 = (-3)$ x (-3) x $(-3) = \boxed{-27}$

3) $3^3 = 3$ x 3 x $3 = 27$ 4) $49 = 7^2$ 5) $-64 = (-4)^3$

EXERCISES

For 1-6, rewrite each as an integer.

  1) 8^2 2) $(-2)^4$ 3) $(-3)^4$

  4) 4^3 5) $(-5)^4$ 6) 2^5

For 7-12, rewrite each as an exponent.

  7) 3 x 3 x 3 x 3 x 3 8) 81   9) 64

 10) (-2) (-2) (-2) 11) 121   12) (-1) (-1) (-1)

REVIEW

Simplify each of the following.

  1) $-\frac{2}{3} \div -\frac{1}{2}$ 2) $15 - 17 + -2$

  3) $\frac{3}{5} + -\frac{1}{2}$ 4) -2.6 x -5

INTRODUCTION

Exponents are used throughout algebra. There are some important laws of exponents that you need to know. It is well worth it to memorize these laws. You will use them a lot all through math.

Helpful • $a^m \times a^n = a^{m+n}$
Hints
• $(\frac{a}{b})^m = \frac{a^m}{b^m}$

• $(a^m)^n = a^{m \times n}$

• $(a \times b)^m = a^m \times b^m$

• $\frac{a^m}{a^n} = a^{m-n}$

• $a^{-n} = \frac{1}{a^n}$

• It is often good to have positive exponents in your answers.

EXAMPLES

1) $5^2 \times 5^3 = 5^{2+3} = 5^5$

2) $(\frac{5}{2})^3 = \frac{5^3}{2^3}$

3) $(2^3)^4 = 2^{3 \times 4} = 2^{12}$

4) $(3 \times 5)^3 = 3^3 \times 5^3$

5) $\frac{2^5}{2^3} = 2^{5-3} = 2^2$

6) $6^{-3} = \frac{1}{6^3}$

7) $5^1 = 5$

8) $9^0 = 1$

9) $\frac{4^2}{4^5} = 4^{2-5} = 4^{-3} = \frac{1}{4^3}$

10) $7^{2-8} = 7^{-6} = \frac{1}{7^6}$

EXERCISES

Simplify each of the following.

1) $5^4 \times 5^7$

2) $(\frac{7}{4})^2$

3) $(2 \times 5)^4$

4) $(4^2)^3$

5) $\frac{2^7}{2^3}$

6) 6^{-2}

7) $3^3 \times 3^2$

8) $(4 \times 5)^2$

9) $(3^2)^3$

10) 2^{-4}

11) $\frac{3^2}{3^4}$

12) $(5^3)^2$

REVIEW

Simplify each of the following.

1) 5^2

2) 3^4

3) $79 - 96$

4) $5 \times -\frac{3}{4}$

1-9 Square Roots

The symbol for **square root** is $\sqrt{}$. The expression $\sqrt{36}$ is read "the square root of 36". The answer is the number which when multiplied by itself is equal to 36. That number is 6. Keeping this in mind, $\sqrt{49} = 7$, because 7 x 7 = 49, and $\sqrt{81} = 9$, because 9 x 9 = 81. There are a few important rules for square roots which are worth memorizing.

Helpful Hints

• $\sqrt{a^2} = a$

• $\sqrt{a \times b} = \sqrt{a} \times \sqrt{b}$

• $\sqrt{\dfrac{a}{b}} = \dfrac{\sqrt{a}}{\sqrt{b}}$

• Numbers such as $\sqrt{2}$, $\sqrt{5}$, and $\sqrt{7}$ are irrational. They cannot be simplified, so leave them as part of the answer.

EXAMPLES

1) $\sqrt{121} = 11$

2) $\sqrt{2500} = \sqrt{25 \times 100} = \sqrt{25} \times \sqrt{100} = 5 \times 10 = 50$

3) $\sqrt{\dfrac{16}{81}} = \dfrac{\sqrt{16}}{\sqrt{81}} = \dfrac{4}{9}$

4) $\sqrt{75} = \sqrt{25 \times 3} = \sqrt{25} \times \sqrt{3} = 5\sqrt{3}$

5) $\sqrt{18} = \sqrt{9 \times 2} = \sqrt{9} \times \sqrt{2} = 3\sqrt{2}$

6) $\sqrt{\dfrac{75}{36}} = \dfrac{\sqrt{75}}{\sqrt{36}} = \dfrac{\sqrt{25 \times 3}}{6} = \dfrac{\sqrt{25} \times \sqrt{3}}{6} = \dfrac{5\sqrt{3}}{6}$

EXERCISES

Simplify each of the following.

1) $\sqrt{25}$

2) $\sqrt{100}$

3) $\sqrt{900}$

4) $\sqrt{400}$

5) $\sqrt{50}$

6) $\sqrt{20}$

7) $\sqrt{\dfrac{81}{9}}$

8) $\sqrt{\dfrac{25}{36}}$

9) $\sqrt{\dfrac{8}{25}}$

10) $\sqrt{3600}$

11) $\sqrt{18}$

12) $\sqrt{\dfrac{72}{8}}$

REVIEW

Simplify each of the following.

1) $(-2)^3$

2) $\dfrac{4 \bullet -6}{-2 \bullet 2}$

3) $-\dfrac{1}{2} - \dfrac{1}{4}$

4) $0.3 - 0.7$

INTRODUCTION

It is necessary to follow the correct **order of operations** when simplifying an expression. This is done to ensure that there is exactly one answer. There are a few important grouping symbols that will be used throughout algebra: **Parentheses ( )**, **Brackets [ ]**, **Braces { }**, and the **Fraction Bar** —. A number next to a grouping symbol means to multiply unless there is another sign present. In the expression 3(4 + 7 - 2) you would get an answer inside the parentheses and multiply it by 3.

Use the following order of operations when simplifying expressions:

Helpful Hints
- **First**, evaluate within grouping symbols. Start with the innermost grouping symbol and work outward. In other words, work from the inside out.
- **Second**, eliminate all exponents and square roots.
- **Third**, multiply and divide in order from left to right.
- **Fourth**, add and subtract.
- Be careful to show all your steps.

EXAMPLES

Simplify each of the following. The bold-typed numbers will help you to see the proper order of operation.

1) $3^2 (3 + 5) + 3$

$= 3^2 \mathbf{(8)} + 3$

$= \mathbf{9} (8) + 3$

$= \mathbf{72} + 3$

$= \boxed{75}$

2) $4 + 12 \times 3 - 8 \div 4$

$= 4 + \mathbf{36} - \mathbf{2}$

$= \mathbf{40} - 2$

$= \boxed{38}$

There are no grouping symbols or exponents or square roots. So start with $\times$ and $\div$.

3) $2 + 4 \{8 - [8 - 2 (3 - 1)] \div 2\}$

$= 2 + 4 \{8 - [8 - 2 \mathbf{(2)}] \div 2\}$

$= 2 + 4 \{8 - [8 - \mathbf{4}] \div 2\}$

$= 2 + 4 \{8 - \mathbf{[4]} \div 2\}$

$= 2 + 4 \{8 - \mathbf{2}\}$

$= 2 + 4 \mathbf{\{6\}}$

$= 2 + \mathbf{24} = \boxed{26}$

4) $\dfrac{5 (8 - 3) - 2^2}{3 + 2 (3^2 - 7)}$

$= \dfrac{5 \mathbf{(5)} - 2^2}{3 + 2 \mathbf{(9 - 7)}}$

$= \dfrac{5 (5) - \mathbf{4}}{3 + 2 \mathbf{(2)}}$

$= \dfrac{\mathbf{25} - 4}{3 + \mathbf{4}}$

$= \dfrac{21}{7} = \boxed{3}$

1-10 Order of Operations

Solve each of the following. Be sure to follow the correct order of operations.

1) $5 + 9 \times 3 - 4$

2) $8 + 3^2 \times 4 - 6$

3) $5^2 + (15 + 3) \div 2$

4) $4(6 + 2) - 5^2$

5) $9 + \{(4 + 5) \times 3\}$

6) $6(-3 + 9) + -4$

7) $4^3 - 7(2 + 3)$

8) $(12 + -3) + 75 \div 5^2$

9) $\dfrac{(12 - 3) + 3^2}{-7 + 2(4 + 1)}$

10) $\dfrac{7^2 - (-5 + 9)}{2(4^2 - 12) - 3}$

11) $6(5 + 3) - 6^2$

12) $4^2(3 + 7) + -6$

13) $12 \div 2 \bullet 4 - 15$

14) $3\{(5 + 4) \bullet 3 - 5\} \div 3$

15) $3(5^2 - 1) - 12$

16) $2 + \{3(6 - 2) \div 2\} - 4$

17) $\dfrac{3^2 (4^2)}{3(5 - 1)}$

18) $2^3 + 5(2 - 3)$

REVIEW

Simplify each of the following.

1) $\sqrt{81}$

2) $\sqrt{\dfrac{16}{25}}$

3) 3^4

4) $\sqrt{27}$

Chapter 1: **NECESSARY TOOLS FOR ALGEBRA** 21

1-11 Properties of Numbers

The **rules** for working with real numbers are based on what we call **properties**. We will accept these properties as facts. For any real numbers, a, b, and c, the following properties are true.

EXAMPLES

Helpful Hints

- Identity Property of Addition — $0 + a = a$ — $0 + 2 = 2$
- Identity Property of Multiplication — $1 \times a = a$ — $1 \times 7 = 7$
- Inverse Property of Addition — $a + (-a) = 0$ — $5 + (-5) = 0$
- Inverse property of Multiplication — $a \times \frac{1}{a} = 1 \ (a \neq 0)$ — $6 \times \frac{1}{6} = 1$
- Associative Property of Addition — $(a+b) + c = a + (b + c)$ — $(2 + 3) + 4 = 2 + (3 + 4)$
- Associative Property of Multiplication — $(a \times b) \times c = a \times (b \times c)$ — $(2 \times 3) \times 4 = 2 \times (3 \times 4)$
- Commutative Property of Addition — $a + b = b + a$ — $5 + 6 = 6 + 5$
- Commutative Property of Multiplication — $a \times b = b \times a$ — $4 \times 3 = 3 \times 4$
- Distributive Property — $a \times (b + c) = a \times b + a \times c$ — $5 \times (3 + 2) = 5 \times 3 + 5 \times 2$

EXERCISES

Name the property that is illustrated.

1) $7 + 9 = 9 + 7$

2) $3 \times (7 + 4) = 3 \times 7 + 3 \times 4$

3) $7 + (-7) = 0$

4) $3 \times (4 \times 5) = (3 \times 4) \times 5$

5) $0 + (-6) = -6$

6) $5 \times \frac{1}{5} = 1$

7) $9 + (6 + 5) = 9 + (5 + 6)$

8) $9 \times 7 = 7 \times 9$

9) $(6+5) + 7 = 6 + (5+7)$

10) $1 \times \frac{7}{8} = \frac{7}{8}$

11) $3(2) + 3(4) = 3(2 + 4)$

12) $16 + -16 = 0$

REVIEW

Simplify each of the following.

1) $6 - -\frac{1}{2}$

2) $3(-2)(-3)$

3) $6 + 4(6 + 2)$

4) $3 + 15 \div 3 - 4$

1-12 The Number Line

INTRODUCTION

A **number line** allows us to "picture" numbers. There are a few important facts that you should know about number lines.

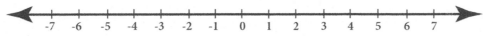

Helpful Hints

- Positive numbers are located to the right of zero.
- Negative numbers are located to the left of zero.
- Numbers are graphed on a number line with a point.
- The farther to the right of zero, the greater the value of a number. The farther to the left of zero, the lesser the value of a number.

EXAMPLES

Numbers can be assigned to a point on a number line. Positive numbers are to the right of zero. Negative numbers are to the left of zero.

Numbers are graphed on a number line with a point.

Examples: A is the graph of -5.   A has a coordinate of -5.

               B is the graph of -1.   B has a coordinate of -1.

               C is the graph of 5.   C has a coordinate of 5.

               D is the graph of $7\frac{1}{2}$.   D has a coordinate of $7\frac{1}{2}$.

EXERCISES

Use the number line to state the coordinates of the given points.

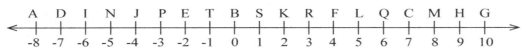

1) B   2) D, E, and G   3) L and H

4) R and F   5) K, F, and C   6) N and A

7) G, H, I, and Q   8) H, D, and S   9) A, M, B, and P

10) B, C, and M   11) L, F, and P   12) L, P, H, and A

REVIEW

Simplify each of the following.

1) $-9 - 15 =$ 2) $-17 + -6 - 11 =$

3) $3(-7)(-2) =$ 4) $-355 \div 5 =$

1-13 The Coordinate Plane

INTRODUCTION

A **coordinate plane** is used to locate or plot points, lines, and various types of graphs. They are a very important part of algebra. There are some important facts you need to know about a coordinate plane.

Helpful Hints

- The **horizontal** line is called the **x-axis** and the **vertical** line is called the **y-axis**.
- The point where the x-axis and y-axis intersect is called the **origin**.
- The first number of an ordered pair (x, y) shows how to move across. It is called the **x-coordinate**. The second number of an order pair shows how to move up or down. It is called the **y-coordinate**.
- The x-coordinate is called the **abscissa**, and the y-coordinate is called the **ordinate**.
- A coordinate plane is divided into four sections called **quadrants**. They are labeled with Roman numerals.

EXAMPLES

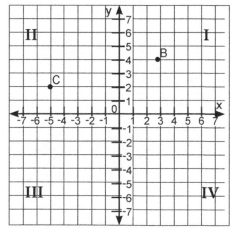

Examples (left): **To locate B, move across to the right to 3 and up 4. The ordered pair is (3,4).**

To locate C, move across to the left to -5 and up 2. The ordered pair is (-5,2).

Examples (right): **(-5, 3) is found by moving across to the left to -5, and up 3. This is represented by point B. -5 is the x-coordinate and 3 is the y-coordinate.**

(6, 3) is found by moving across to the right to 6, and up 3. This is represented by point C. 6 is the x-coordinate and 3 is the y-coordinate.

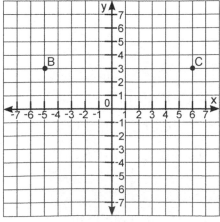

Chapter 1: **NECESSARY TOOLS FOR ALGEBRA**

1-13 The Coordinate Plane

EXERCISES

Use the coordinate system to find the ordered pair associated with each point.

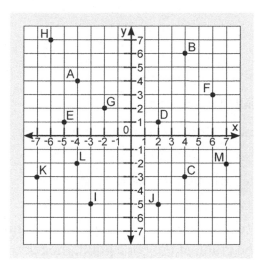

Part A

1) D	2) L
3) F	4) J
5) K	6) E
7) B	8) C
9) I	10) G
11) D	12) H

Use the coordinate system to find the point associated with each ordered pair.

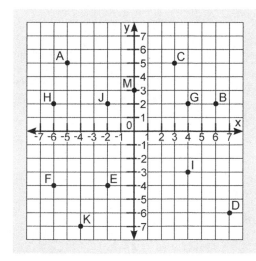

Part B

1) (6, 2)	2) (-5, 5)
3) (3, 5)	4) (7, -6)
5) (-6, -4)	6) (0, 3)
7) (-2, -4)	8) (-2, 2)
9) (-6, 2)	10) (4, 2)
11) (-4, -7)	12) (4, -3)

REVIEW

Simplify each of the following.

1) $-9 - 6$

2) $-\frac{2}{3} \div \frac{1}{6}$

3) $\frac{\sqrt{100}}{\sqrt{4}}$

4) $3\{2(5 + -7)\}$

1-14 Relations and Functions

A **relation** and a **function** are similar, but it is important to be able to tell them apart. You need to know these facts about relations and functions.

Helpful Hints
- A **relation** is a set of ordered pairs.
- The x-values in a set of ordered pairs are called the **domain**.
- The y-values in a set of ordered pairs are called the **range**.
- A **function** is a special relation in which each value of x is paired with exactly one value of y.
- Graphing a relation makes it very easy to tell whether or not the relation is a function.
- If a vertical line can intersect the graph at two or more points, the relation is **not** a function. If a vertical line can intersect the graph at only one point, the relation **is** a function. This is called the **vertical line test**.

EXAMPLES

R = { (2,3), (3,4), (4,5), (6,7) }

The domain is {2, 3, 4, 6} (all of the x-values)

The range is {3,4,5,7} (all of the y-values)

R **is** a function because each x-value is paired with exactly one y-value.

Here is the graph of the function. Notice that a vertical line can intersect the graph at only one point.

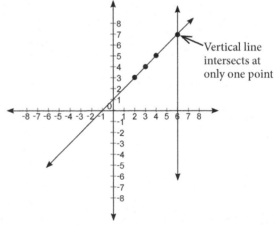

Vertical line intersects at only one point

D = { (0,3), (2,1), (3,0), (2,-2) }

The domain is {0,2,3} (all the x-values)

The range is {-2,0,1,3} (all the y-values)

D is **not** a function because 2, an x-value, is paired with two y-values, 1 and -2.

Here is the graph of the relation. Notice that a vertical line can intersect the graph at two points.

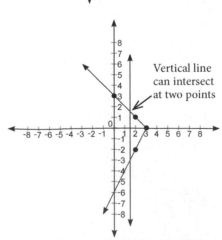

Vertical line can intersect at two points

1-14 Relations and Functions

EXERCISES

1) Is F = {(2,3), (2,2), (6,7)} a function? Why?

2) Is M = {(3,4), (2,4), (5,5)} a function? Why?

3) What is the domain of F?

4) What is the term for the x and y coordinates?

5) Explain the vertical line test.

Use R = {(1,2),(2,3),(5,6),(9,7)} to answer questions 6–8.

6) List the domain of R.

7) List the range of R.

8) Is R a function? Why?

Use T = {(1,2),(3,2),(3,4),(5,7)} to answer questions 9–11.

9) List the domain of T.

10) List the range of T.

11) Is T a function? Why?

REVIEW

Simplify each of the following.

1) $\frac{1}{3} + -\frac{1}{2}$

2) $-\frac{1}{5} + -\frac{1}{2}$

3) $30 \div 6 \times 2 + 7$

4) $2.3 + -6.5$

1-15 Factors, Divisibility Tests, and Prime Factorization

Whole numbers that we multiply are called **factors** of their **product**. For example, in the equation **2 x 3 = 6**, we see that 2 and 3 are the factors, and 6 is the product. A **prime number** is any whole number greater than one whose only factors are one and itself. For example, 19 is a prime number since no whole number will divide into it evenly except for 1 and 19. It is often necessary to find the **prime factors** of a number. A prime factor is just a prime number that divides evenly into a whole number with no remainder. A **composite** number is any whole number greater than 1 that is **not** prime.

Divisibility tests are useful in finding factors without actually dividing. Here is a list of divisibility tests that are worth memorizing.

Helpful Hints

- **2** is a factor of any whole number that ends in 0, 2, 4, 6, or 8.
- **3** is a factor of a whole number if the sum of its digits is divisible by 3. For example, **4,251** is divisible by 3 since 4+2+5+1=12, and 12 is divisible by 3.
- **4** is a factor of a whole number if the last two digits are divisible by 4. For example, **3,432** is divisible by 4 since 32 (the last two digits) is divisible by 4.
- **5** is a factor of a whole number that ends in 0 or 5.
- **6** is a factor of a whole number if both 2 and 3 are factors.
- **8** is a factor of a whole number if the last three digits are divisible by 8. For example, **4,320** is divisible by 8 since 320 (the last three digits) is divisible by 8.
- **9** is a factor of a whole number if the sum of the digits is divisible by 9. For example **3,456** is divisible by 9 since 3+4+5+6=18, and 18 is divisible by 9.
- **10** is a factor of a whole number that ends in 0.
- **Prime factorization** means to write a number as a product of its prime factors. This can be done using the **shortcut division method**.

EXAMPLES

1) **Test 23,340 for divisibility by 2, 3, 4, 5, 6, 8, 9, and 10.**

   First, the sum of the digits is 12.
   - 2 **is** a factor since the number is even.
   - 3 **is** a factor since 3 is a factor of 12 which is the sum of the digits.
   - 4 **is** a factor since 4 is a factor of the last two digits 40.
   - 5 **is** a factor since the number ends in 0.
   - 6 **is** a factor since both 2 and 3 are factors.
   - 8 **is not** a factor since it is not a factor of the last three digits 340.
   - 10 is a factor since the number ends in 0.

   So, the factors of **23,340** are 2, 3, 4, 5, 6, and 10.

1-15 Factors, Divisibility Tests, and Prime Factorization

Remember, **prime factorization** means to write a number as a product of its prime factors. Use shortcut division to find the prime factorization of each number. Express your answer using exponents.

36

2	36	Divide by 2
2	18	Divide by 2
3	9	Divide by 3
	3	Finished

$36 = 2 \times 2 \times 3 \times 3$
$ = 2^2 \times 3^2$

140

2	140	Divide by 2
2	70	Divide by 2
5	35	Divide by 5
	7	Finished

$140 = 2 \times 2 \times 5 \times 7$
$ = 2^2 \times 5 \times 7$

420

2	420	Divide by 2
2	210	Divide by 2
3	105	Divide by 3
5	35	Divide by 5
	7	Finished

$420 = 2 \times 2 \times 3 \times 5 \times 7$
$ = 2^2 \times 3 \times 5 \times 7$

EXERCISES

Test each of the following for divisibility by 2, 3, 4, 5, 6, 8, 9, and 10.

1) 346

2) 620

3) 3,420

4) 8,244

Use shortcut division to find the prime factorization of each of the following. Express your answer using exponents.

5) 90

6) 48

7) 112

8) 100

9) 84

10) 300

11) 320

12) 96

13) 312

14) 120

15) 220

16) 450

REVIEW

Simplify each of the following.

1) $7.6 + -9.3$

2) $3^2 \times 3^3$

3) $\dfrac{2^5}{2^3}$

4) 7^0

1-16 Greatest Common Factors

INTRODUCTION

The **greatest common factor** is the largest number that is a factor of 2 two or more given numbers. For example the greatest common factor of **12** and **9** is equal to 3, since 3 is the largest number that divides evenly into both 9 and 12. The greatest common factor is often abbreviated as **GCF**. There is a simple technique to find the greatest common factor that works well with larger numbers.

Helpful
Hints
- **First**, use shortcut division to find the prime factorization of the given numbers.
- **Second**, identify the common prime factors. These are the highest power of each factor tht appears in **all** of the prime factorizations.
- **Third**, the GCF is the product of all the common prime factors
- **Remember**, the answer represents the **largest** number that will divide evenly into each of the given numbers.

EXAMPLES

Find the GCF of 120 and 90.

2	120		2	90	*Shortcut*
2	60		3	45	*Division*
2	30		3	15	
3	15			5	
	5				

$120 = ②\text{ x } 2 \text{ x } 2 \text{ x} ③\text{x}⑤$ *Circle the factors common to both numbers.*

$90 = ②\text{x}③\text{x } 3 \text{ x}⑤$

GCF = 2 x 3 x 5 = 30 *The GCF is the product of all the common prime factors.*

Find the GCF of 45, 63, and 60

3	45		3	63		2	60	*Shortcut*
3	15		3	21		2	30	*Division*
	5			7		3	15	
							5	

$45 = ③\text{x } 3 \text{ x } 5$ *Circle the factors common to both numbers.*

$63 = ③\text{x } 3 \text{ x } 7$

$60 = 2 \text{ x } 2 \text{ x}③\text{x } 5$

GCF = 3 *The GCF is the product of all the common prime factors.*

EXERCISES

Find the greatest common factor for each of the following.

1) 48, 16   2) 36, 40   3) 72, 80

4) 112, 120   5) 130, 200   6) 150, 96

7) 24, 30, 48   8) 42, 56, 70   9) 45, 70, 84

REVIEW

Simplify each of the following.

1) 7.9 + -6.2

2) $-\frac{1}{4} + -\frac{1}{4}$

3) -3 {72 ÷ 8 + 5}

4) $\sqrt{63}$

1-17 Least Common Multiples

INTRODUCTION

A **least common multiple** is the smallest number that two or more numbers can divide into evenly with no remainder. For example, the least common multiple of **6** and **9** is 18, since 18 is the **smallest** number that both 6 and 9 can divide into evenly. The least common multiple is often abbreviated as **LCM**. Finding the least common multiple is like finding the **lowest common denominator** when adding or subtracting fractions. There is a simple technique to find the least common multiple of two or more numbers.

Helpful Hints
- **First,** use shortcut division to find the prime factorization of the given numbers. Be sure to express the prime factorizations using exponents.
- **Second,** identify the highest power of each factor that appears in **any** of the prime factorizations. The highest power needs to appear only once. It can appear more than once.
- **Third,** the **LCM** is the product of these factors.
- Remember, the answer represents the smallest number that **all** of the given numbers can divide into evenly.

EXAMPLES

Find the LCM of 16 and 18

$$
\begin{array}{r|r}
2 & 16 \\ \hline
2 & 8 \\ \hline
2 & 4 \\ \hline
 & 2
\end{array}
\qquad
\begin{array}{r|r}
2 & 18 \\ \hline
3 & 9 \\ \hline
 & 3
\end{array}
$$

$16 = 2^4$
$18 = 2 \times 3^2$

Find the highest power of each factor.

$\text{LCM} = 2^4 \times 3^2$
$\quad = 16 \times 9 = 144$

Find the LCM of 12, 10, and 8

$$
\begin{array}{r|r}
2 & 12 \\ \hline
2 & 6 \\ \hline
 & 3
\end{array}
\qquad
\begin{array}{r|r}
2 & 10 \\ \hline
 & 5
\end{array}
\qquad
\begin{array}{r|r}
2 & 8 \\ \hline
2 & 4 \\ \hline
 & 2
\end{array}
$$

$12 = 2^2 \times 3$
$10 = 2 \times 5$
$8 = 2^3$

Find the highest power of each factor.

$\text{LCM} = 2^3 \times 3 \times 5$
$\quad = 8 \times 3 \times 5 = 120$

EXERCISES

Find the least common multiple for each of the following.

1) 12, 20
2) 18, 20
3) 16, 20
4) 21, 12
5) 6, 8, 12
6) 10, 18, 20
7) 7, 14, 21
8) 40, 36
9) 54, 90

REVIEW

Simplify each of the following.

1) $(-3)^3$
2) $\sqrt{27}$
3) $-.79 - .67$
4) $(-2)^4$

INTRODUCTION

Scientific notation is used to express very large and very small numbers. Scientific notation is quite useful in science. A number in scientific notation is expressed as the **product of two factors**. The **first** factor is a number between 1 and 10 and the **second** factor is **power of 10**. For example, **2.36×10^5** and **3.976×10^{-7}** are numbers which are both expressed in scientific notation. Remember the following.

Helpful Hints
- Large and small number can be changed to scientific notation.
- Numbers expressed in scientific notation can be changed back to conventional numbers.

EXAMPLES

1. **Change 157,000,000,000 to scientific notation.**

   Simply move the decimal between the 1 and 5. Since the decimal has moved 11 places to the **left**, the answer is **1.57×10^{11}**.

2. **Change .0000000468 to scientific notation.**

   Move the decimal between the 4 and 6. Since the decimal has moved eight places to the **right**, the answer is **4.68×10^{-8}**.

3. **Change 3.458×10^8 to a conventional number.**

   Move the decimal eight places to the **right**. The answer is **345,800,000**.

4. **Change 4.5677×10^{-7} to a conventional number.**

   Move the decimal seven spaces to the **left**. The answer is **.00000045677**.

EXERCISES

Change each of the following numbers to scientific notation.

1) 12,360,000,000

2) .000000149

3) 159,700

4) .000007216

1-18 Scientific Notation

5) 1,096,000,000   6) .0001963

7) .0000000016   8) 7,900,000,000

9) .000093   10) 93,000,000

Change each number in scientific notation to a conventional number.

11) 7.032×10^5 12) 2.3×10^4

13) 5×10^{-5} 14) 1.127×10^4

15) 2.1×10^4 16) 2.1×10^{-3}

17) 3.2×10^5 18) 7×10^6

19) 2.61×10^{-3} 20) 3.56×10^5

REVIEW

1) Simplify $-7 - 9 - 6$ 2) Simplify $3 \cdot 6 \, (-2)$

3) Find the GCF of 36 and 80.   4) Find the LCM of 12 and 20.

1-19 Ratios and Proportions

INTRODUCTION

A **ratio** compares two numbers or groups of objects. Notice in the diagram below that for every three circles there are four squares. The ratio can be written in following ways: **3 to 4**, **3:4**, and $\frac{3}{4}$. Each of these is read as "three to four".

In algebra, ratios are often written in fraction form. Remember that ratios expressed as fractions can be reduced to lowest terms.

Two equal ratios can be written as a **proportion**. The following is an example of a proportion: $\frac{4}{6} = \frac{2}{3}$

In a proportion, the **cross products** are equal. For example, to determine whether $\frac{3}{4} = \frac{5}{6}$ is a proportion, simply **cross multiply**. $\frac{3}{4} \bowtie \frac{5}{6}$

$3 \times 6 = 18, \quad 4 \times 5 = 20, \quad 18 \neq 20 \quad$ It **is not** a proportion.

To determine whether $\frac{6}{9} = \frac{8}{12}$ is a proportion, again we can cross multiply. $\frac{6}{9} \bowtie \frac{8}{12}$

$6 \times 12 = 72, \quad 9 \times 8 = 72, \quad 72 = 72 \quad$ It **is** a proportion.

It is easy to find the **missing number** in a proportion. For example, in the proportion $\frac{4}{n} = \frac{2}{3}$ we can easily find the value of n.

$\frac{4}{n} \bowtie \frac{2}{3}$

First, cross multiply: $2 \times n = 4 \times 3$

$$2 \times n = 12$$

Next, divide 12 by 2: $2 \overline{)12}^{\,6}$ $\boxed{n = 6}$

Remember the following when solving proportions.

Helpful Hints
- A ratio compares two numbers or groups of objects.
- A proportion is formed when two ratios are equal.
- In a proportion, the cross products are equal.
- To find the missing number in a proportion, cross multiply.

EXAMPLES

1) **Write the ratio 36 to 9 as a fractions reduced to its lowest terms.**

$$\frac{36}{9} = \frac{4}{1}$$

2) **Cross multiply to determine whether $\frac{5}{3} = \frac{14}{9}$ is a proportion.**

$\frac{5}{3} \bowtie \frac{14}{9}$ $3 \times 14 = 42, \quad 5 \times 9 = 45, \quad 43 \neq 45 \quad$ It **is not** a proportion.

Chapter 1: **NECESSARY TOOLS FOR ALGEBRA**

1-19 Ratios and Proportions

3) **Find the missing number in the following proportion.**

$\frac{n}{4} = \frac{12}{16}$ Cross Multiply $16 \times n = 4 \times 12$

$16 \times n = 48$ Divide 48 by 16.

$16\overline{)48}$ → 3 $\boxed{n = 3}$

4) **Find the missing number in the following proportion.**

$\frac{4}{5} = \frac{x}{7}$ Cross Multiply $5 \times x = 4 \times 7$

$5 \times x = 28$ Divide 28 by 5.

$5\overline{)28}$ → $5\frac{3}{5}$, -25, 3 $\boxed{x = 5\frac{3}{5}}$

EXERCISES

Write each of the following ratios as fractions reduced to lowest terms.

1) 5 bats to 3 balls   2) 16:12

3) 30 books to 25 pens   4) 30 to 45

Cross multiply to determine whether each of the following is a proportion.

5) $\frac{2}{5} = \frac{6}{15}$ 6) $\frac{5}{2} = \frac{11}{4}$

7) $\frac{15}{20} = \frac{6}{8}$ 8) $\frac{3}{1.3} = \frac{9}{3.5}$

Find the missing number in each proportion.

9) $\frac{3}{15} = \frac{n}{5}$ 10) $\frac{n}{40} = \frac{5}{100}$

11) $\frac{7}{n} = \frac{3}{9}$ 12) $\frac{21}{7} = \frac{7}{n}$

13) $\frac{n}{2} = \frac{7}{5}$ 14) $\frac{27}{3} = \frac{n}{2}$

15) $\frac{n}{5} = \frac{7}{3}$ 16) $\frac{\frac{1}{2}}{\frac{1}{3}} = \frac{\frac{1}{4}}{n}$

REVIEW

Use T = {(0,1), (2,3), (0,6), (7,4)} to answer 1-3.

1) Is set T a function?   2) List the domain.

3) List the range.   4) Simplify $\frac{4^7}{4^5}$

1-20 Using Proportions in Word Problems

INTRODUCTION

Proportions can often be used to solve word problems. All you need to do is set up a proportion and find the missing number. It is important to be consistent when you set the proportion up. For example, if you indicate hours at the top of one ratio, be sure that hours is on the top of the other ratio. Just remember to follow the following steps.

Helpful Hints
- **First**, set up a proportion. Label the missing part with a variable (letter).
- **Second**, cross multiply to find the answer
- Remember to be consistent. For example, if hours is on the bottom of one ratio, it has to be on the bottom of the other.
- Label your answer with a word or a short phrase.

EXAMPLE

Ratios and **proportions** can be used to solve problems.

Example: A car can travel 384 km in six hours. How far can the car travel in eight hours?

First set up a proportion. $\dfrac{384 \text{ km}}{6 \text{ hours}} = \dfrac{n \text{ km}}{8 \text{ hours}}$ → Next, divide by six.

Next, cross multiply. $6 \times n = 8 \times 384$

$6 \times n = 3{,}072$

$$6 \overline{\smash{)}3072} \quad 512$$

$\boxed{n = 512}$

The car can travel 512 km in eight hours.

EXERCISES

Use a proportion to solve each problem.

1) A car can travel 100 km on five liters of gas. How many liters will be needed to travel 40 km?

2) Two kg of chicken cost $7. How much will five kg cost?

3) In a class, the ratio of boys to girls is four to three. If there are 20 boys in the class, how many girls are there?

4) A hiker takes three hours to go 24 km. At this rate, how far could he hike in five hours?

5) Seven kg of nuts cost $5. How many kg of nuts can you buy with $2?

6) If 3 apples cost $1.19, how much would 18 apples cost?

7) A man can travel 13 km on his bicycle in 2 hours. At this rate, how far can he travel in 5 hours?

8) On a map, 1 cm = 30 km. How many cm on the map would represent 135 km?

9) A man earned $180 for working 14 hours. How many hours must he work to earn $300?

10) In a class, the ratio of boys to girls is 8 to 5. If there are 75 girls in the class, how many boys are there in the class?

11) The weight of 80 meters of wire is 5 kg. What is the weight of 360 meters of wire?

12) The owner of a house worth $72,000 pays $2,000 in taxes. At this rate, how much will the taxes be on a house worth $54,000?

13) A rectangular photo measures 15 cm by 20cm. It is enlarged so that the length of the shorter side is 21 cm. What would be the length of the longer side?

14) If 12 oranges cost $4.80, what would be the cost of 5 oranges?

REVIEW

Simiplify each of the following.

1) -.16 + .2

2) 2.1 – 3.7

3) 3 x -2.6

4) $\frac{-2.13}{-3}$

1-21 Percents

INTRODUCTION

Percent means "**per hundred**" or "**hundredths**". The symbol for percent is **%**. Percents can be expressed as **decimals** and **fractions**. The fraction form may sometime be reduced to its lowest terms. You will find that percent is used a lot in everyday life. The following examples show the close relationship between percents, fractions, and decimals.

$$25\% = .25 = \frac{25}{100} = \frac{1}{4} \qquad 8\% = .08 = \frac{8}{100} = \frac{2}{25}$$

Remember the following.

Helpful Hints
- When working with percents you may work with fractions or decimals. Use the one that is the easiest.

- There are three basic types of percent problems. Learn to recognize them.

EXAMPLES

Type I Percent Problems: Finding **the percent of a number.** You may use either fractions or decimals, whatever is most convenient.

Find 25% of 60. $.25 \times 60$

$$\begin{array}{r} 60 \\ \times .25 \\ \hline 300 \\ 120 \\ \hline \underline{15.00} \end{array}$$

OR

Since $25\% = \frac{25}{100} = \frac{1}{4}$

$$\frac{1}{4_1} \times \frac{\overset{15}{\cancel{60}}}{1} = \frac{15}{1} = \boxed{15}$$

Type II Percent Problems: Finding the **percent.** The answer will have a **percent sign (%)** in the answer. **First,** write a fraction. **Second,** change the fraction to a decimal. **Third,** change the decimal to a percent.

4 is what percent of 16?

$$\frac{4}{16} = \frac{1}{4}$$

$$\begin{array}{r} .25 = \boxed{25\%} \\ 4\overline{)1.00} \\ \underline{-\ 8\downarrow} \\ 20 \\ \underline{-20} \\ 0 \end{array}$$

5 is what percent of 25?

$$\frac{5}{25} = \frac{1}{5}$$

$$\begin{array}{r} .20 = \boxed{20\%} \\ 5\overline{)1.00} \\ \underline{-\ 10} \\ 00 \end{array}$$

Type III Percent Problems: Finding the **whole** when the **part** and the **percent** are known. Simply change the equal sign (=) to the division sign (÷).

6 = 25% of what number?
$6 \div 25\%$ (Change = to ÷.)
$6 \div .25$ (Change % to decimal.)

$$.25\overline{)6.00}\ \ \overset{24.}{}$$

Twelve is 80% of what number?
$12 \div 80\%$ (Change = to ÷.)
$12 \div .8$ (Change % to decimal.)

$$.8\overline{)12.0}\ \ \overset{15.}{}$$

* Be careful to move decimal points properly.

1-21 Percents

EXERCISES

Solve each of the following. Remember that you can use fractions or decimals. Pick the one that is easiest to use.

For 1-4, change each percent to a decimal and to a fraction reduced to its lowest terms.

1) 20%

2) 9%

3) 45%

4) 75%

Solve each of the following.

5) Find 70% of 25.

6) Find 6% of 72.

7) Find 25% of 80.

8) Find 90% of 200.

9) 15 is what % of 60?

10) 24 is what % of 32?

11) 13 = what % of 60?

12) 16 is 20% of what?

13) 8 is 40% of what?

14) Find 70% of 200.

15) Find 7% of 80.

16) 24 is what % of 40?

17) Find 22% of 60.

18) Find 75% of 32.

REVIEW

Simplify each of the following.

1) $-\frac{1}{2} + -\frac{1}{3}$

2) $-\frac{1}{2} - \frac{7}{8}$

3) $-\frac{1}{2} \times -1\frac{1}{2}$

4) $-3\frac{1}{2} \div \frac{1}{2}$

INTRODUCTION

What good is math if you can't put it to good use? Percents are used often in a variety of everyday situations. Remember that there are basically three types of percent problems, and all you need to be able to do is identify them. Remember the following tips.

Helpful Hints
- Type I problems ask you to find the **percent of a number**.
- Type II problems ask you to find the **percent**. In other words, there will be a percent sign (%) in the answer.
- Type III problems ask you to find the **whole** when the part and the percent are known.

EXAMPLES

Type I

A man earns $300 and spends 40% of it. How much does he spend? (You know the whole.)

Find 40% of 300.

$$\begin{array}{r} \$300 \\ \times\ .4 \\ \hline \$120 \end{array}$$

He spends $120.

Type II

In a class of 25 students, 15 of them are girls. What percent are girls?
(Your answer will have a percent sign—%.)

15 = what % of 25?

$$\frac{15}{25} = \frac{3}{5} \qquad 5\overline{)3.00} = .60 = 60\%$$

60% of the class are girls.

Type III

Five students got A's on a test. This is 20% of the class. How many students are there in the class? (You are finding the whole.)

5 = 20% of the class

5 ÷ .2

$$.2\overline{)5.0} = 25.$$

There are 25 students in the class.

1-22 Percent Word Problems

EXERCISES

1) On a test with 25 questions, Al got 80% correct. How many questions did he get correct?

2) A player took 12 shots and made 9. What percent did he make?

3) A girl spent $5. This was 20% of her earnings. How much were her earnings?

4) Buying a $8,000 car requires a 20% down payment. How much is the down payment?

5) 3 = 10% of what?

6) A team played 20 games and won 18. What % did they win?

7) A farmer sold 50 cows. If this was 20% of his herd, how many cows were in his herd?

8) 20 = 80% of what?

9) Paul wants a bike that costs $400. If he has saved 60% of this amount, how much has he saved?

10) There are 400 students in a school. Fifty-five percent are girls. How many boys are there?

11) 12 is what % of 60?

12) Kelly earned 300 dollars and put 70% of it into the bank. How much did she put into the bank?

REVIEW

Simplify each of the following.

1) $-\frac{1}{2} + \frac{3}{4} =$

2) $-\frac{1}{3} - \frac{1}{4} =$

3) $-\frac{1}{2} \times 3 =$

4) $-2\frac{1}{2} \div -1\frac{1}{4} =$

Chapter 1 Review: Necessary Tools for Algebra

1) -6 + 7 + -9

2) -6 – -6 – 9

3) 3(-2) (-3)

4) -36 ÷ 9

5) $\frac{1}{5} + -\frac{3}{4}$

6) $\frac{2}{5} - \frac{4}{5}$

7) $1\frac{1}{2} \cdot -1\frac{1}{3}$

8) $\dfrac{\frac{3}{5}}{\frac{1}{4}}$

9) -2.5 + -3.6

10) 1.4 – 3.6

11) 2.3 x -1.7

12) -.333 ÷ -0.3

13) $(-3)^4$

14) $2^2 \cdot 2^4$

15) $\dfrac{7^8}{7^6}$

16) $\sqrt{400}$

17) 6 + 7 x 6 – 5

18) $3\{(6 - 2) + 3^2\}$

Chapter 1 Review: Necessary Tools for Algebra

For 19–21, R = {(1, 2), (2, 7), (3, 2), (4, 8)}

19) List the domain.

20) List the range.

21) Is R a function?

22) Find the prime factorization for 108.

23) Find the Greatest Common Factor for 120 and 180.

24) Find the Least Common Multiple for 24 and 15.

25) Write 321,000,000 in scientific notation.

26) Write .00000362 in scientific notation.

27) Write 2.73×10^{-4} in standard form.

28) Solve the proportion: $\frac{1}{2} = \frac{x}{5}$

29) Solve the proportion: $\frac{3}{n} = \frac{15}{45}$

30) Find 15% of 90.

31) 6 is what % of 30?

32) 12 = 25% of what?

33) In a class of 45 students, 80% received A's on a Math test. How many students received an A?

34) Mary spent 15 dollars. If this was 20% of her earnings, how much money did she earn?

2-1 Solving Equations Using Addition and Subtraction

INTRODUCTION

An **equation** is a math sentence that contains an equal sign (=) and states that one expression is equal to another. In algebra, equations will contain a **variable**, which is a letter that stands for one or more unknown numbers. To solve an equation, the variable should be isolated (separated by itself) on one side of an equal sign, and the answer (solution) on the other. An example of this is x = 7. **It is good to have the variable on the left side of the equation if possible.**

When solving an equation you can add, subtract, multiply, or divide as long as you do the same on each side of the equal sign. For example, if 2 is added to the expression on the left side of the equal sign, then 2 must be added to the expression on the right side of the equal sign. In other words, each side of the equal sign has to be treated exactly the same.

Here is some important information about equations that you need to remember.

Helpful Hints
- An **equation** is a math sentence that contains an equal sign (=) and states that one expression is equal to another.
- A **variable** is a letter used to represent one or more numbers.
- A **term** is a part of an expression separated by + or – signs. In the expression 3x – 2y + 3z, the terms are 3x, 2y, and 3z.
- A **constant** is a number or quantity that does not change. In the expression x + 5, 5 is a constant.
- A **coefficient** is a number that multiplies a variable. For example, in 5x, 5 is the coefficient.
- A **numerical expression** involves only numbers. For example, 4 + 7 is a numerical expression.
- An **algebraic expression** is a numerical expression that includes variables. For example, 3x + 5 is an algebraic expression.
- **When solving an equation, remember this. If you add or subtract a number on one side of the equal sign, you must add or subtract the same number on the other side of the equal sign.**

EXAMPLES

Solve each of the following equations. **Check your answers by substituting them back into the original equation.**

1) $x + 3 = 17$

$x + 3 - 3 = 17 - 3$ *Subtract 3 from both sides.*

$x = 14$

2) $x + 5 = -12$

$x + 5 - 5 = -12 - 5$ *Subtract 5 from both sides.*

$x = -17$

Check your answers.

2-1 Solving Equations Using Addition and Subtraction

3) $5 = x - 7$

  $x - 7 = 5$ *Rewrite with the variable on the left.*

  $x - 7 + 7 = 5 + 7$ *Add 7 to both sides.*

  $x = 12$

4) $n - 6 = -12$

  $n - 6 + 6 = -12 + 6$ *Add 6 to both sides.*

  $n = -6$

Check your answers.

EXERCISES

Solve each of the following equations. Check your answers by substituting them back into the original equation.

1) $n + 3 = 12$

2) $x - 7 = 15$

3) $x + 3 = -6$

4) $n - 9 = -6$

5) $15 = x - 2$

6) $23 = x + 6$

7) $-7 = n - 3$

8) $-8 = x + 6$

9) $x - 3 = 76$

10) $x - 9 = -26$

11) $-27 = x + -6$

12) $n - -2 = 15$

13) $y - 3 = 16$

14) $n + 8 = 29$

15) $72 + x = 89$

16) $n - -6 = -7$

17) $-9 = x - 7$

18) $-76 = 43 + x$

REVIEW

1) Find 6% of 120.

2) Find 60% of 120.

1) 3 is what % of 15?

2) 30 is 20% of what?

2-2 Solving Equations Using Multiplication and Division

INTRODUCTION

Solving equations using multiplication and division is similar to using addition and subtraction. Just remember that if you multiply or divide by a number on one side of the equal sign, you must multiply or divide by the same number on the other side of the equal sign. Both sides of the equal sign must be treated the same. Remember the following when using multiplication and division.

Helpful Hints
- A number in front of a variable indicates multiplication. The equation $3n = 15$ means "3 times n is equal to 15."
- An equation like the example $\frac{x}{3} = 4$ indicates division. The equation means "x divided by 3 is equal to 4."
- It is good to have the variable on the left side of the equation if possible.
- Be careful of negative signs.
- **When solving an equation, remember this: If you multiply or divide by a number on one side of the equal sign, you must multiply or divide by the same number on the other side of the equal sign.**

EXAMPLES

Solve each of the following equations. Check your answers by substituting them into the original equations.

1) $2x = 16$

$\frac{2x}{2} = \frac{16}{2}$ *Divide both sides by 2.*

$x = 8$

2) $\frac{x}{3} = 9$

$3 \cdot \frac{x}{3} = 3 \cdot 9$ *Multiply both sides by 3.*

$x = 27$

3) $-5x = -45$

$\frac{-5x}{-5} = \frac{-45}{-5}$ *Divide both sides by -5.*

$x = 9$

4) $\frac{x}{-4} = 12$

$-4 \cdot \frac{x}{-4} = -4 \cdot 12$ *Multiply both sides by -4.*

$x = -48$

5) $\frac{-x}{2} = 6$

$2 \cdot \frac{-x}{2} = 6 \cdot 2$ *Multiply both sides by 2.*

$-x = 12$

$x = -12$

6) $\frac{1}{2}x = 10$

$\frac{\frac{1}{2}x}{\frac{1}{2}} = \frac{10}{\frac{1}{2}}$ *Divide each side by $\frac{1}{2}$.*

(Remember to invert $\frac{1}{2}$.)

$x = 20$

2-2 Solving Equations Using Multiplication and Division

EXERCISES

Solve each equation. Check your answers by substituting them back into the original equation.

1) $3x = 24$

2) $\frac{x}{7} = 9$

3) $\frac{x}{2} = -4$

4) $3x = -12$

5) $-2x = -40$

6) $\frac{n}{-3} = 7$

7) $-12x = -24$

8) $\frac{n}{-2} = -5$

9) $15x = 75$

10) $\frac{n}{-10} = 6$

11) $\frac{x}{3} = 15$

12) $3x = -63$

13) $\frac{-x}{4} = 5$

14) $\frac{-n}{2} = -6$

15) $\frac{1}{2}n = 14$

16) $\frac{1}{3}n = -5$

17) $1\frac{1}{2}x = 6$

18) $12x = -36$

REVIEW

Solve each proportion.

1) $\frac{3}{5} = \frac{x}{15}$

2) $\frac{5}{4} = \frac{30}{x}$

3) $\frac{2}{9} = \frac{6}{x}$

4) $\frac{3}{x} = \frac{5}{6}$

2-3 Solving 2-Step Equations

INTRODUCTION

You know how to solve equations that require just one step. Two-step equations are easy. Just remember, our goal is to isolate the variable on one side of the equal sign, and the answer on the other. It is good to have the variable on the left side. Remember the following when solving 2-step equations.

Helpful Hints
- There are only two steps
- **First**, if necessary, rewrite the equation with the variable on the left side of the equal sign.
- **Second**, just like in the one-step equations, perform the addition, subtraction, multiplication, or division to isolate the variable by itself.
- Remember to treat each side of the equal sign the same.
- It is good to have the variable on the left of the equal sign.
- Be careful of negative signs.

EXAMPLES

Solve each equation. Check your answers by substituting them back into the original equation.

1) $3x - 5 = 16$

$3x - 5 + 5 = 16 + 5$ *Add 5 to both sides.*

$3x = 21$

$\dfrac{3x}{3} = \dfrac{21}{3}$ *Divide both sides by 3.*

$x = 7$

2) $7x + 3 = -25$

$7x + 3 - 3 = -25 - 3$ *Subtract 3 from both sides.*

$7x = -28$

$\dfrac{7x}{7} = \dfrac{-28}{7}$ *Divide both sides by 7.*

$x = -4$

3) $\dfrac{x}{5} - 6 = 9$

$\dfrac{x}{5} - 6 + 6 = 9 + 6$ *Add 6 to both sides.*

$\dfrac{x}{5} = 15$

$5 \bullet \dfrac{x}{5} = 5 \bullet 15$ *Multiply both sides by 5.*

$x = 75$

4) $-6x + 2 = -28$

$-6x + 2 - 2 = -28 - 2$ *Subtract 2 from both sides.*

$-6x = -30$

$\dfrac{-6x}{-6} = \dfrac{-30}{-6}$ *Divide both sides by -6.*

$x = 5$

     Chapter 2: **SOLVING EQUATIONS**

2-3 Solving 2-Step Equations

Solve each equation. Check your answers by substituting them back into the original equation.

1) $3x - 5 = 16$

2) $7x + 3 = -4$

3) $\frac{x}{2} + 2 = 4$

4) $-14n - 7 = 49$

5) $2n + 45 = 15$

6) $\frac{n}{5} + -6 = 9$

7) $4x - 10 = 38$

8) $-2m + 9 = 7$

9) $35n + 12 = 82$

10) $\frac{x}{-3} - 6 = -10$

11) $-40 = 2x - 10$

12) $2x + 1 = -15$

13) $\frac{n}{3} + 4 = 13$

14) $\frac{x}{2} - 7 = 65$

15) $\frac{n}{5} + 6 = 24$

16) $-5x + 15 = 45$

17) $\frac{x}{3} - 7 = -3$

18) $7x + 12 = -2$

REVIEW

Simplify each of the following.

1) $\frac{1}{6} + -\frac{1}{2}$

2) $-\frac{1}{5} - -\frac{2}{3}$

3) $\frac{2}{3} \bullet -1\frac{1}{2}$

4) $-2\frac{1}{2} \div \frac{1}{2}$

INTRODUCTION

Sometimes there are variables on both sides of the equal sign. To solve these, **isolate the variable on one side**. It is good to have the variable on the left side of the equal sign. It doesn't have to be on the left, but in general, this is a good way to go in many cases. While learning algebra it is good to develop a routine when you approach equations. Remember the following when solving equations with variables on both sides of the equal sign.

Helpful Hints
- The first step is to get the variable on the left side of the equal sign. So start by removing the variable located on the right side of the equal sign.
- If you add, subtract, multiply, or divide on one side of the equal sign, you must do the same on the other side of the equal sign.
- Be careful with negative signs.
- Check your answers by substituting them back into the original equation.

EXAMPLES

Solve each of the following equations. Check your answers by substituting them back into the original equation.

1)
$$5x - 6 = 2x + 9$$
$$5x - 2x - 6 = 2x - 2x + 9 \quad \text{\textit{Substract 2x from both sides.}}$$
$$3x - 6 = 9$$
$$3x - 6 + 6 = 9 + 6 \quad \text{\textit{Add 6 to both sides.}}$$
$$3x = 15$$
$$\frac{3x}{3} = \frac{15}{3} \quad \text{\textit{Divide both sides by 3.}}$$
$$x = 5$$

2)
$$-6x + 12 = 4x - 8$$
$$-6x - 4x + 12 = 4x - 4x - 8 \quad \text{\textit{Substract 4x from both sides.}}$$
$$-10x + 12 = -8$$
$$-10x + 12 - 12 = -8 - 12 \quad \text{\textit{Substract 12 from both sides.}}$$
$$-10x = -20$$
$$\frac{-10x}{-10} = \frac{-20}{-10} \quad \text{\textit{Divide both sides by -10.}}$$
$$x = 2$$

EXERCISES

Solve each of the following equations. Check your answers.

1) $3x + 6 = x + 8$

2) $6n - 5 = 4n + 9$

3) $4x + 4 = 2x + 6$

4) $7n - 3 = 3n + 5$

5) $3r + 7 = 10r + 28$

6) $8x - 8 = -4x + 16$

7) $3m - 15 = 7m + 5$

8) $4x + 5 = 6x + 7$

9) $3x - 8 = -4x + 13$

10) $9x - 13 = -5x + 71$

11) $3x + 5 = 2x + 16$

12) $7x - 4 = 2x + 6$

13) $8n = 9 - n$

14) $7x + 10 = 3x + 50$

15) $2x + 36 = -3x - 54$

16) $5x - 13 = 43 - 2x$

17) $14 - 8x = 3x - 8$

18) $n + 30 = 12n - 14$

REVIEW

1) Give the prime factorization for 210.

2) Find 12% of 80.

3) Write .000009 in scientific notation.

4) Simplify $3 + 16 \div 2 \times 3 - 6$

2-5 Solving Equations Using the Distributive Property

Sometimes the **distributive property** can be used to solve equations. The distributive property is often used to get rid of parentheses.

Helpful Hints
- The distributive property states: $a \times (b + c) = a \times b + a \times c$
- Use the distributive property to get rid of the parentheses.
- Once the parentheses are gone, the problem is often quite easy to solve using the skills that you have learned.

EXAMPLES

Solve the following. Check your answers by substituting them back into the original equation.

1) $2(x + 7) = 30$

Use the distributive property

$2(x + 7) = 30$

$2x + 14 = 30$

$2x + 14 - 14 = 30 - 14$ *Subtract 14 from both sides.*

$2x = 16$

$\dfrac{2x}{2} = \dfrac{16}{2}$ *Divide both sides by 2.*

$x = 8$

2) $3(4x - 3) = -33$

Use the distributive property

$3(4x - 3) = -33$

$12x - 9 = -33$

$12x - 9 + 9 = -33 + 9$ *Add 9 to both sides.*

$12x = -24$

$\dfrac{12x}{12} = \dfrac{-24}{12}$ *Divide both sides by 12.*

$x = -2$

3) $-4(x - 2) = -12$

Use the distributive property

$-4(x - 2) = -12$

$-4x + 8 = -12$

$-4x + 8 - 8 = -12 - 8$ *Subtract 8 from both sides.*

$-4x = -20$

$\dfrac{-4x}{-4} = \dfrac{-20}{-4}$ *Divide both sides by -4.*

$x = 5$

4) $6(x - 1) = 3(x + 1)$

Use the distributive property

$6(x - 1) = 3(x + 1)$

$6x - 6 = 3x + 3$

$6x - 3x - 6 = 3x - 3x + 3$ *Subtract 3x from both sides.*

$3x - 6 = 3$

$3x - 6 + 6 = 3 + 6$ *Add 6 to both sides.*

$3x = 9$

$\dfrac{3x}{3} = \dfrac{9}{3}$ *Divide both sides by 3.*

$x = 3$

2-5 Solving Equations Using the Distributive Property

Use the distributive property to solve each of the equations. Check your answers.

1) $2(n + 3) = 10$

2) $2(x - 2) = 6$

3) $-4(x + 2) = 16$

4) $2(2x + 6) = 24$

5) $-5(x + 2) = 30$

6) $-5(y - 7) = -20$

7) $4(m - 1) = 2m$

8) $5(x - 3) = 2x + 3$

9) $5(m + 4) = 2m + 20$

10) $2(x + 3) = 3(x - 3)$

11) $8(n - 1) = 4(n + 4)$

12) $7(2n + 3) = 2(6n + 9)$

13) $3(x - 5) + 19 = -2$

14) $x + 4(x - 2) = -1$

15) $2(x - 2) + 28 = 2(2x + 1)$

16) $14x + 8 = 5(x - 3) + 5$

REVIEW

1) Simplify $3^2 \times 5^2$

2) Simplify $\frac{5^9}{5^7}$

3) Solve
   $3m + 2 = 11$

4) Solve
   $2x - 9 = 3$

2-6 Solving Equations by Collecting Like Terms

INTRODUCTION

It is important to be able to recognize and combine **like terms** when solving equations. Like terms have the same variables to the same power. For example, 5m and 3m are like terms. Another example of a pair of like terms is 3xy and 8xy. Notice that the coefficients do not have to be the same. The terms $3a^2b$ and 2ab are **not** like because they don't have the same variables to the same power. To collect like terms just add or subtract the coefficients. For example, 7xy + 2xy = 9xy, and 3a – 7a = -4a.

Helpful Hints
- When solving an equation and you recognize like terms, combine them.
- To combine the like terms just add or subtract the coefficients.
- In many equations it is necessary to **remove parentheses** using the **distributive property**, and them combine like terms.

EXAMPLES

Solve the following. Check your answers by substituting them back into the original equation.

1) $10x + 4 - 4x = -20$

$6x + 4 = -20$ *Combine like terms.*

$6x + 4 - 4 = -20 - 4$ *Subtract 4 from both sides.*

$6x = -24$

$\dfrac{6x}{6} = \dfrac{-24}{6}$ *Divide both sides by 6.*

$x = -4$

2) $7x + 2x - 7 = 21 + 8$

$9x - 7 = 29$ *Combine like terms.*

$9x - 7 + 7 = 29 + 7$ *Add 7 to both sides.*

$9x = 36$

$\dfrac{9x}{9} = \dfrac{36}{9}$ *Divide both sides by 9.*

$x = 4$

3) $7x - 4 - 3x = 2x + 10$

$4x - 4 = 2x + 10$ *Combine like terms.*

$4x - 2x - 4 = 2x - 2x + 10$

$2x - 4 = 10$

$2x - 4 + 4 = 10 + 4$ *Add 4 to both sides.*

$2x = 14$

$\dfrac{2x}{2} = \dfrac{14}{2}$ *Divide both sides by 2.*

$x = 7$

4) $7(x + 2) - 4x = 2(x + 5)$

$7(x + 2) - 4x = 2(x + 5)$ *Use distributive properties to remove parentheses.*

$7x + 14 - 4x = 2x + 10$

$3x + 14 = 2x + 10$ *Combine like terms.*

$3x - 2x + 14 = 2x - 2x + 10$ *Subtract 2x from both sides.*

$x + 14 = 10$

$x + 14 - 14 = 10 - 14$ *Subtract 14 from both sides.*

$x = -4$

2-6 Solving Equations by Collecting Like Terms

Solve each equation. Remember, if necessary remove parentheses using the distributive property. Check your answers by substituting them back into the original equation.

1) $9x - 16 - 3x = 20$

2) $7x + 16 - 11x = 28$

3) $3a + 6a - 7 + 2 = 24 - 2$

4) $7x - 5x = 24 - 8$

5) $14 - 2x + 4x = 80 - 8$

6) $7n + 2 + 3n = 12$

7) $16 = 3x + 2x - 9$

8) $4 (n + 3) = 6n + 8 + 2n$

9) $5 (x + 4) = 2x + 20 + 2x$

10) $3 (x - 5) = 2x + 5 + 3x$

11) $2 (x + 4) = 2 (x - 4) + 4x$

12) $3 (y + 2) = 3 (y - 2) + 3y$

13) $26 = 3x + 2x - 9$

14) $9x - 3x - x = 95$

15) $5x + 5x + 3 = 4 (2x + 1) - 2$

16) $3 (5x - 2) + 4 = 11x - 3 - 4x$

Solve each of the following.

1) $\frac{m}{3} - 4 = 1$

2) $2x - 7 = -25$

3) $3 (x + 5) = 36$

4) $7 (x + 2) = -35$

2-7 Solving Equations Involving Absolute Value

INTRODUCTION

The **absolute value** of a number is its distance from 0 on a number line. The absolute value of a number is **never** negative. The expression $|$-5$|$ is read "**the absolute value of -5.**"

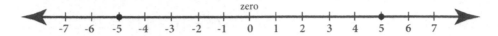

The absolute value of -5 = 5 because -5 is 5 units from 0 on the number line. The absolute value of 5 = 5 because 5 is also 5 units from 0 on the number line. It is often necessary to solve equations involving absolute value. Remember the following tips when solving equations involving absolute value.

Helpful Hints
- The symbol for absolute value is $|$ $|$.
- Absolute value is a number's distance from 0 on a number line.
- An absolute value is never negative.
- Often an equation involving absolute values will have two solutions.

EXAMPLES

Solve the following. Check your answers by substituting them back into the original equation.

1) $3|x| + 5 = 8$

$3|x| + 5 - 5 = 8 - 5$ *Subtract 5 from both sides.*

$3|x| = 3$

$\dfrac{3|x|}{3} = \dfrac{3}{3}$ *Divide both sides by 3.*

$|x| = 1$

$x = 1$ or $x = -1$

2) $6|x| - 9 = 15$

$6|x| - 9 + 9 = 15 + 9$ *Add 9 to both sides.*

$6|x| = 24$

$\dfrac{6|x|}{6} = \dfrac{24}{6}$ *Divide both sides by 6.*

$|x| = 4$

$x = 4$ or $x = -4$

3) $|2x + 4| = 18$

$2x + 4 = 18$ — OR — $2x + 4 = -18$

$2x + 4 - 4 = 18 - 4$ *Subtract 4 from both sides.* $2x + 4 - 4 = -18 - 4$ *Subtract 4 from both sides.*

$2x = 14$ $2x = -22$

$\dfrac{2x}{2} = \dfrac{14}{2}$ *Divide both sides by 2.* $\dfrac{2x}{2} = \dfrac{-22}{2}$ *Divide both sides by 2.*

$x = 7$ $x = -11$

$x = 7$ or $x = -11$

2-7 Solving Equations Involving Absolute Value

Solve each of following equations. Check your answers by substituting them back into the original equation.

1) $3|x| + 7 = 28$

2) $6|x| - 5 = 31$

3) $2|x| - 7 = -3$

4) $7|n| + 13 = 34$

5) $|x + 8| = 10$

6) $|2x - 7| = 17$

7) $|x - 12| = 16$

8) $|3x - 10| = 5$

9) $|x - 2| = 4$

10) $4|x| - 1 = 15$

11) $2|y| - 2 = 7$

12) $|3x - 2| = 10$

13) $|2x + 1| = 13$

14) $4|4 - x| = 16$

15) $|6 - 2x| = 2$

16) $|3x + 8| = 29$

REVIEW

Solve each equation.

1) $\frac{x}{7} = -9$

2) $2(2m + 6) = 24$

3) $4x + 4 = 2x + 6$

4) $8m - 5 = 3m + 25$

2-8 Simplifying Algebraic Expressions Containing Parentheses

INTRODUCTION

Working with **parentheses** can sometimes be confusing. It is important to know how to work with expressions contained in parentheses. In many problems, it is necessary to remove parentheses to find solutions.

When removing parentheses many students often have difficulties with positive and negative signs especially when working with subtraction. Remember the following important facts. They can prevent a lot of mistakes.

Helpful Hints
- When removing parentheses with a **plus sign (+)** in front, get rid of the plus sign and the parentheses and **do not change** the signs of any of the terms. Remember to collect like terms.

- When removing parentheses with a **minus sign (−)** in front, get rid of the minus sign and the parentheses and **change** the sign of each of the terms. Remember to collect like terms.

EXAMPLES

1) $(3x − 2xy + 3) + (7x + 5xy − 7)$

   $= 3x − 2xy + 3 + 7x + 5xy − 7$ *Remove parentheses and get rid of plus sign (+). Do not change any signs.*

   $= 3x + 7x − 2xy + 5xy + 3 − 7$ *Collect like terms and simplify.*

   $= 10x + 3xy − 4$

2) $(5y + 2xy + 3) + (-6y + 3xy + 4)$

   $= 5y + 2xy + 3 − 6y + 3xy + 4$ *Remove parentheses and get rid of plus sign (+). Do not change any signs.*

   $= 5y − 6y + 2xy + 3xy + 3 + 4$ *Collect like terms and simplify.*

   $= -y + 5xy + 7$

3) $(4x − 2xy + 5) − (3x + 8xy + 7)$

   $= 4x − 2xy + 5 − 3x − 8xy − 7$ *Remove parentheses and get rid of minus sign (−). Change the sign of each term.*

   $= 4x − 3x − 2xy − 8xy + 5 − 7$ *Collect like terms and simplify.*

   $= x − 10xy − 2$

4) $(3x + 2xy − 3) − (-5x + 3xy − 8)$

   $= 3x + 2xy − 3 + 5x − 3xy + 8)$ *Remove parentheses and get rid of minus sign (−). Change the sign of each term.*

   $= 3x + 5x + 2xy − 3xy − 3 + 8$ *Collect like terms and simplify.*

   $= 8x − xy + 5$

2-8 Simplifying Algebraic Expressions Containing Parentheses

EXERCISES

Simplify each of the following.

1) $(5x + 7y) + (-2x - 2y)$

2) $(8x + 4y) - (2x + 7y)$

3) $(6m - 3x) - (-4m - 6x)$

4) $(-7x + 8y) + (7x - 11y)$

5) $(2y + 3x - 7) + (3y - 4x - 4)$

6) $(3m - 4n + 8) - (6m + 8n + 3)$

7) $(-5x + 3y - 7) - (-8x - 3y - 4)$

8) $(3x - 4xy + 7) + (-11x + 3y + 7)$

9) $(3x - 9y) - (-7x - 8y)$

10) $(7x + 9y) + (-12x - 15y)$

11) $(5x - 7y) - (12x + 2y)$

12) $(7x - 2xy + y) + (-9x + 7y)$

13) $(3x - y + 2z) - (-2x - 8y - 3z)$

14) $(-3x^2 + 2y - 3x^2y) + (2x^2 - 3y + 2x^2y)$

15) $(6x - y + 3) - (2x + 7y - 9)$

16) $(7x^2 - 5x) - (6x - 3x^2 + xy)$

REVIEW

Simiplify each of the following.

1) $\frac{7^3}{7^2}$

2) $n^7 \cdot n^8$

3) $(n^2)^3$

4) $\sqrt{121}$

2-9 Solving Multi-Step Equations

INTRODUCTION

Some equations take several steps. Often there is more than one effective strategy to use. Remember to use all the algebra tools that you have learned. Also remember that if you add, subtract, multiply, or divide on one side of the equal sign, you must do the same on the other side of the equal sign. It is also important to work neatly, carefully, and avoid making careless errors. Remember the following tips as you work through multi-step equations.

Helpful Hints
- It is good to isolate the variable on the left if possible.
- Remember to treat each side of the equal sign the same.
- Be careful with negative signs.
- Use the distributive law to get rid of parentheses when possible.
- Combine like terms.
- If there is division by a number, eliminate it by multiplying both sides of the equal sign by that number.
- Check your answers by substituting them back into the original equation.

EXAMPLES

Solve each of the following equations. Check your answers by substituting them back into the original equation.

1) $7(x-2) - 6 = 2x + 8 + x$

First use distributive property and collect like terms.

$$7x - 14 - 6 = 2x + 8 + x$$
$$7x - 20 = 3x + 8$$
$$7x - 3x - 20 = 3x - 3x + 8 \quad \textit{Subtract 3x from both sides.}$$
$$4x - 20 = 8$$
$$4x - 20 + 20 = 8 + 20 \quad \textit{Add 20 to both sides.}$$
$$4x = 28$$
$$\frac{4x}{4} = \frac{28}{4} \quad \textit{Divide both sides by 4.}$$
$$x = 7$$

2) $\dfrac{x+4}{5} = x$

$$5 \cdot \frac{x+4}{5} = 5 \cdot x \quad \textit{Multiply both sides by 5.}$$
$$x + 4 = 5x$$
$$x - 5x + 4 = 5x - 5x \quad \textit{Subtract 5x from both sides.}$$
$$-4x + 4 = 0$$
$$-4x + 4 - 4 = 0 - 4 \quad \textit{Subtract 4 from both sides.}$$
$$-4x = -4$$
$$\frac{-4x}{-4} = \frac{-4}{-4} \quad \textit{Divide both sides by -4.}$$
$$x = 1$$

3)
$$\frac{x-2}{2} = 5$$

$$2 \cdot \frac{(x-2)}{2} = 2 \cdot 5 \quad \textit{Multiply both sides by 2.}$$

$$x - 2 = 10$$

$$x - 2 + 2 = 10 + 2 \quad \textit{Add 2 to both sides.}$$

$$x = 12$$

4)
$$x = \frac{x}{2} + 5$$

$$2 \cdot x = 2\left(\frac{x}{2} + 5\right) \quad \textit{Multiply both sides by 2. Use distributive property.}$$

$$2x = x + 10$$

$$2x - x = x - x + 10 \quad \textit{Subtract x from both sides.}$$

$$x = 10$$

EXERCISES

Solve each of the following equations. Check your answers by substituting them back into the original equation.

1) $7(x + 1) = 5(x - 3)$

2) $3(x - 3) - 4x = 5$

3) $4(x - 3) - x = x - 6$

4) $\frac{2x + 4}{2} = 2$

5) $\frac{6x - 1}{2} = 4$

6) $\frac{7x + 3}{3} = 2x - 1$

7) $\frac{x}{2} - 3 = 7$

8) $\frac{5x - 4}{2} = 18$

9) $\frac{x - 8}{3} = -10$

10) $7(m + 2) - 4m = 2(m + 10)$

11) $\frac{x + 40}{15} = -4$

12) $\frac{15 + 9x}{6} = 7$

13) $-9 - 3(2x - 1) = -18$

14) $\frac{2}{3}n + 6 = 16$

REVIEW

Solve each equation.

1) $-m + 4m = 15$

2) $7x - 2 - 5x = 0$

3) $3(x - 2) = 2x$

4) $4(x + 3) = 6x + 8$

Chapter 2 Review: Solving Equations

Solve each equation. Be sure to show your steps. Check your answers by substituting back into the original equation.

1) $x + 12 = 36$

2) $x - 7 = -19$

3) $5n = 115$

4) $\frac{n}{3} = -7$

5) $7x - 3 = 32$

6) $\frac{x}{5} + 5 = -2$

7) $5x + 8 = 3x + 20$

8) $6n - 5 = 8n - 25$

9) $5(4n + 2) = 100$

10) $2(m+2) = 3m + 2$

11) $5y = 4y - 42 - 3y$

12) $32 = 6x + 4x - 18$

13) $|x + 3| = 10$

14) $12|x| - 18 = 30$

15) $\frac{4(x + 16)}{2} = 24$

16) $7x - 26 = 5(2 - x)$

17) $3(4x - 9) = 5(2x - 5)$

18) $\frac{3x}{2} - 9 = 0$

Don't Forget to Use the Free Online Video Lessons

www.nononsensealgebra.com

Go to www.nononsensealgebra.com and you will find a corresponding video lesson for each lesson in the book. The author, award-winning teacher, Richard W. Fisher, will carefully guide you through each topic, step-by-step. Each lesson will provide easy-to-understand instruction, and the student can work examples right along with Mr. Fisher. It's like having your own personal math tutor available 24/7. After the video lesson, you can go to the book and complete the lesson. The book combined with the video lessons will turbo-charge your ability to master algebra.

**Your access code
to the online lessons is located
on page 1 of this book.**

3-1 Graphing Linear Equations

INTRODUCTION

The graph of a **linear equation** is always a straight line. In a linear equation there are **two variables**, each to the **first power**. The standard form of a linear equation is **ax + by = c**, where a, b, and c are integers and a and b are not both zero. A linear equation has an infinite number of solutions, so to make a graph, select a few points and graph them. Then we draw a line that connects the points. Graphing a linear equation is a simple process.

Helpful Hints
- **First**, select four values for x.
- **Second**, find the values for y by substituting each x-value into the original equation.
- **Third**, plot the points and connect them with a line.
- To make an accurate graph, use graph paper when doing your work.

EXAMPLES

The graph of a **linear equation** is always a line. A linear equation can have an infinite number of solutions, so to make a graph we select a few points and graph them, and then draw a line that connects them.

Draw a graph of the solutions to the following equation.

$$y = x + 3$$

First, select four values for x and find the values for y. Start with x = 0 and make a chart like the one to the right.

x	y	
0	3	(0,3)
1	4	(1,4)
2	5	(2,5)
4	7	(4,7)

Next, plot the points and connect them with a line.

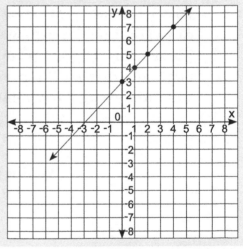

Draw a graph of the solutions to the following equation.

$$y = \frac{x}{2} + 2$$

* Select values for x which are multiples of 2.

x	y	
0	2	(0,2)
2	3	(2,3)
4	4	(4,4)
-2	1	(-2,1)

Next, plot the points and connect them with a line.

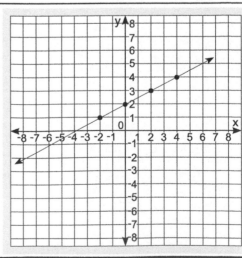

3-1 Graphing Linear Equations

Graph each of the following linear equations. Select 4 values for x and find the value for y. Graph and connect the points. Use graph paper.

1) $y = 2x + 1$

2) $y = 2x - 1$

3) $y = \frac{x}{2}$

4) $y = \frac{x}{3}$

5) $y = \frac{x}{2} + 5$

6) $y = -2x$

7) $y = -x + 2$

8) $y = 2x + 8$

9) $y = \frac{1}{2}x + 2$

10) $y = -2x + 5$

11) $y = 2x + 2$

12) $y = -x + 3$

13) $y = 3x - 2$

14) $y = x - 5$

REVIEW

Solve each equation.

1) $|x + 4| = 10$

2) $|x - 4| = 10$

3) $6|x| - 9 = 21$

4) $3|x| + 7 = 28$

3-2 Graphing Linear Equations Using x and y Intercepts

INTRODUCTION

It is possible to graph a linear equation by finding the **x and y intercepts**. The **x-intercept** is where a line intersects the **x-axis**, and the **y-intercept** is where the same line intersects the **y-axis**. Once the intercepts are found, simply connect them with a line. Sometimes an equation is in one variable and the graph will be a vertical or horizontal line. Finding the x and y intercepts is a simple process.

Helpful Hints
- **First**, to find the x-intercept, let y = 0, and solve for x.
- **Second**, to find the y-intercept, let x = 0, and solve for y.
- **Third**, connect the two intercepts with a line.
- Use graph paper when doing your work.

EXAMPLES

Graph each of the following.

1) **Graph x = 3**

   The graph is a vertical line.

2) **Graph y = 2**

   The graph is a horizontal line.

3) **Find the x and y intercepts for the equation 2x + 4y = 8, and draw the graph.**

   To find the x-intercept, let y = 0, and solve for x.

$$2x + 4y = 8$$
$$2x + 4(0) = 8$$
$$2x = 8$$
$$x = 4$$

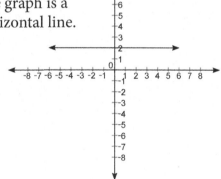

   To find the y-intercept, let x = 0, and solve for y.

$$2x + 4y = 8$$
$$2(0) + 4y = 8$$
$$4y = 8$$
$$y = 2$$

EXERCISES

Find the x and y intercepts for each equation and draw the graph.

1) $6x + 2y = 18$ 2) $3x + 4y = 12$

3) $2x + y = 3$ 4) $2x + 3y = 6$

5) $x + 3y = 9$ 6) $x + y = 5$

7) $x - 3y = 9$ 8) $-4x + y = 2$

9) $-2x + 5y = 10$ 10) $5x + 2y = 10$

11) $x + y = 3$ 12) $3x - 6y = 18$

13) $2x + 3y = 6$ 14) $y + 3x = 6$

15) $2x + y = 8$ 16) $5x = 4y + 20$

REVIEW

Solve each of the following.

1) $5 = 20\%$ of what number?   2) Find 20% of 400.

3) Find 2% of 400.   4) $30 =$ what $\%$ of 120?

3-3 Slope of a Line

The **slope** of a line refers to how steep the line is. Slope is the ratio of the **rise** to the **run**. The change in the **y** direction is called the **rise**. The change in the **x** direction is called the **run**. If you know any two points on a line it is possible to find the slope.

The formula to find the slope of a line is   **slope** $= \frac{y_2 - y_1}{x_2 - x_1}$

Here are some important facts that you need to know about the slop of a line.

Helpful Hints
- The slope of a **horizontal** line is 0.
- The slope of a **vertical** line is **undefined** because the denominator would be 0.
- If the slope is **positive**, the line inclines upward from left to right.
- If the slope is **negative**, the line inclines downward from left to right.
- **Parallel** lines have the same slope.
- **Perpendicular** lines have slopes whose product is -1. For example, $-3 \cdot \frac{1}{3} = -1$, so lines with these slopes would be perpendicular.

EXAMPLE

The slope of a line refers to how steep the line is. It is the ratio of **rise** to **run**.

$$\text{slope} = \frac{y_2 - y_1}{x_2 - x_1}$$

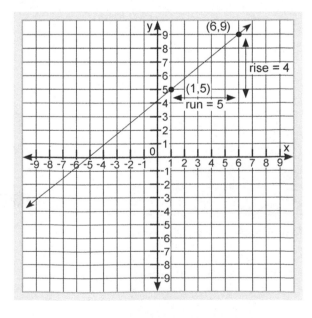

What is the slope of the line passing through the ordered pairs (1, 5) and (6, 9)?

$$\text{slope} = \frac{y_2 - y_1}{x_2 - x_1} \quad \overset{x_1 \ y_1}{(1, 5)}, \overset{x_2 \ y_2}{(6, 9)}$$

$$= \frac{9 - 5}{6 - 1}$$

$$= \left(\frac{4}{5}\right)$$

The **run** is 5 and the **rise** is 4.

3-3 Slope of a Line

EXERCISES

Find the slope of each line that passes through the given points.

1) (2, 3), (5, 4)

2) (3, -2), (5, 1)

3) (4, 3), (2, 6)

4) (4, 1) (7, 2)

5) (-2, 1), (-3, 3)

6) (-2, -2), (6, 3)

7) (4, 5), (6, 6)

8) (1, 2), (3, 9)

9) (1, -1), (6, 5)

10) (3, 2), (8, 6)

11) (2, -1), (4, 2)

12) (9, 2), (7, 5)

13) (-3, 5), (-1, 6)

14) (8, 1), (4, -4)

15) (3, 6), (-3, 4)

16) (-7, -4), (3, 3)

17) (2, 6), (-5, 4)

18) (9, -6), (-7, -8)

REVIEW

For each equation, select 4 values for x, and solve for y. Then draw the graph.

1) $y = 2x + 4$

2) $y = \frac{x}{2} + 3$

3) $y = \frac{x}{3} + 1$

4) $y = -x + 1$

3-4 Changing from Standard Form to the Slope-Intercept Form

Remember, the **standard form** of a linear equation is **ax + by = c**.
However, the **slope-intercept form** is often very useful in graphing linear equations.

The slope-intercept form of a linear equation is **y = mx + b**.
In this equation, **m is the slope of the line**, and **b is the y-intercept**.

An equation in the slope-intercept form is easy to graph. Point **b** is where the graph passes through the **y-axis**, and using the **slope** it is easy to graph another point. Then simply connect the two points with a line.

To change an equation in the standard form to the slope-intercept form, simply **solve for y**. Remember these important tips.

Helpful Hints
- To change from standard form to slope-intercept form simply solve for y.
- In the slope-intercept form **y = mx + b**, **m** = slope **b** = y-intercept.
- When the equation is in the slope-intercept form, use the following order to draw the graph.
    First, plot a point at the y-intercept.
    Second, use the slope to plot another point.
    Third, connect the two points with a line.
- A negative slope such as $-\frac{2}{3}$ can be also written $\frac{-2}{3}$ or $\frac{2}{-3}$. It all works out the same.

EXAMPLE

Determine the slope-intercept form of the following linear equation. Then draw a graph of the line.

2x − 3y = 9   *First, solve for y.*

$-3y = -2x + 9$

$\frac{-3y}{-3} = \frac{-2x}{-3} + \frac{9}{-3}$

$y = \frac{2}{3}x - 3$

Slope   *y-intercept*

$\text{slope} = m = \frac{2}{3} \quad \frac{\text{rise}\uparrow}{\text{run}\rightarrow}$

$\text{y-intercept} = -3$

3-4 Changing from Standard Form to the Slope-Intercept Form

Change each of the following linear equations from the standard form to the slope intercept form and graph the line. Use graph paper.   **y = mx + b**

Be careful of negative slopes. **Examples:** $-\frac{3}{4} = \frac{-3}{4} = \frac{3}{-4}$ The graph will be the same.

1) $y + 3x = 7$ 2) $-2x + 5y = -5$

3) $2y + 4x = -8$ 4) $4x + 4y = 2$

5) $4x + y = 3$ 6) $2y - 6x = 2$

7) $3x - 3y = 9$ 8) $-2x + 5y = -5$

9) $2y - 6x = 1$ 10) $x - 4y = -8$

11) $2x + y = 2$ 12) $y - 2x = -5$

13) $4x + 2y = 6$ 14) $y - 3x = 4$

15) $2x + y = -3$ 16) $x + y = -2$

REVIEW

Find the x and y intercept for each of the following.

1) $y = x + 1$ 2) $y = x - 2$

3) $y = 2x$ 4) $y = 2x - 1$

3-5 Determining the Slope-Intercept Equation of a Line

The slope-intercept form of an equation of a line is $y = mx + b$

Sometimes it is necessary to find the **slope-intercept equation** of a line when only certain information is known. Once the slope-intercept equation is determined it is also easy to change that equation to **standard form**. There are two situations that you will often have when determining the slope-intercept equation of a line.

Helpful Hints
- The first is when you know the **slope** and a **point** on the line.
- The second is when you know **two points** on the line.

EXAMPLES

1) **Find the equation of a line in slope-intercept form having slope 4, and passing through the point (3,1). Graph the line.**

   slope = m = 4

   so, y = 4x + b

   To find b (the y-intercept), substitute (3, 1) into the equation.

   y = 4x + b

   1 = 4(3) + b

   1 = 12 + b

   -11 = b   *(the y-intercept)*

   $\boxed{y = 4x - 11}$

   We now have an equation in the slope-intercept form.

   slope = 4   y-intercept = -11

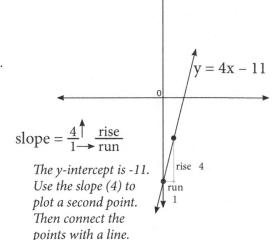

   slope = $\frac{4\uparrow}{1\rightarrow}$ $\frac{\text{rise}}{\text{run}}$

   The y-intercept is -11. Use the slope (4) to plot a second point. Then connect the points with a line.

2) **Find the equation of a line in slope-intercept form passing through the points (2, 1) and (-1, -3). Graph the line.**

   First, find the slope.   **slope** = $\frac{y_2 - y_1}{x_2 - x_2}$

   slope = $\frac{-3-1}{-1-2} = \frac{-4}{-3} = \frac{4}{3}$

   $y = \frac{4}{3}x + b$ *slope-intercept form*

   Substitute either point into the equation. The answer will be the same. Let's use (2, 1).

   $1 = \frac{4}{3} \cdot 2 + b$

   $1 = \frac{8}{3} + b$

   $\frac{3}{3} - \frac{8}{3} = b$

   $-\frac{5}{3} = b = $ *(y-intercept)*

   $\boxed{y = \frac{4}{3}x - \frac{5}{3}}$

   Slope = $\frac{4}{3}$, y-intercept = $-\frac{5}{3}$

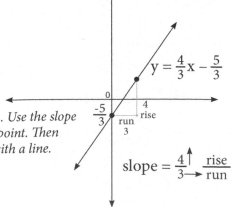

   The y-intercept is $-\frac{5}{3}$. Use the slope ($\frac{4}{3}$) to plot a second point. Then connect the points with a line.

   slope = $\frac{4\uparrow}{3\rightarrow}$ $\frac{\text{rise}}{\text{run}}$

3-5 Determining the Slope-Intercept Equation of a Line

EXERCISES

Find the equation of the line in slope-intercept form. Graph the line.

1) Passes through (2, -3) with slope 4.

2) Passes through (0, 2) with slope 1.

3) Passes through (1, -2) with slope 2.

4) Passes through (-3, 1) with slope 2.

5) Passes through (3, -4) with slope 2.

6) Passes through (-2, 3) and (4, 6).

7) Passes through (2, 1) and (4, 3).

8) Passes through (-2, 3) and (0, 4).

9) Passes through (1, 4) with slope 2.

10) Passes through (1, 4) and (3, 6).

11) Passes through (1, 0) and (3, 6).

12) Passes through (-6, 4) with slope $\frac{2}{3}$.

13) Passes through (3, 1) and (9, 7).

14) Passes through (0, -2) with slope -1.

REVIEW

1) Find 60% of 40.

2) 6 is what % of 30?

3) 3 = 5% of what?

4) What % of 125 is 25?

3-6 Determining the Point-Slope Form

A linear equation in the form $y - y_1 = m(x - x_1)$ is in the **point-slope form**. The slope = m and (x_1, y_1) is a point on the line. The x and y stand for coordinates of a variable point on the line. It is possible to write an equation in this form as long as you know the slope and the coordinates of one point on the line. Remember the following.

Helpful Hints
- To write an equation in the point-slope form you need the slope and a point on the line.
- If you are given just two points, simply find the slope first.

EXAMPLES

1) **Find an equation of a line in point-slope form with slope 4 that passes through the point (1, 3).**

$$y - y_1 = m(x - x_1)$$
$$y - 3 = 4(x - 1) \quad \textit{Point-slope form.}$$

Notice that we can change this to the slope-intercept form.

$$y - 3 = 4(x - 1) \quad \textit{Use the distributive process}$$
$$y - 3 = 4x - 4$$
$$y - 3 + 3 = 4x - 4 + 3 \quad \textit{Add 3 to both sides}$$
$$y = 4x - 1 \quad \textit{Slope-intercept form}$$

2) **Find the equation of a line in point-slope form that passes through the points (4, 5) and (3, -4).**

First, fine the slope.

$$m = \frac{y_2 - y_1}{x_2 - x_1} = \frac{-4 - 5}{3 - 4} = \frac{-9}{-1} = 9$$

$$y - y_1 = m(x - x_1)$$

Substitute either point. The answer will mean the same.

Let's use (4, 5)   — OR —   we could use (3, -4)

$$y - 5 = 9(x - 4) \longleftarrow \text{Point-slope form} \longrightarrow y + 4 = 9(x - 3)$$

For practice, let's change each to the slope-intercept form.

$y - 5 = 9(x - 4)$	$y + 4 = 9(x - 3)$
$y - 5 = 9x - 36$ *Use the distributive process*	$y + 4 = 9x - 27$
$y - 5 + 5 = 9x - 36 + 5$ *Add 5 to both sides*	$y + 4 - 4 = 9x - 27 - 4$ *Subtract 4 from both sides*
$y = 9x - 31 \longleftarrow$ Slope-intercept form	$\longrightarrow y = 9x - 31$

3-6 Determining the Point-Slope Form

Find the equation of a line in the point-slope form for each of the following.

$$y - y_1 = m\,(x - x_1)$$

1) Slope = 5
   passes through (1, 3)

2) Slope = 3
   passes through (1, 2)

3) Slope = 4
   passes through (2, -3)

4) Slope = 3
   passes through (2, 4)

5) Slope = $\frac{3}{4}$
   passes through (-1, 5)

6) Passes through
   (4, 5), (6, 8)

7) Passes through
   (-1, 2), (5, -4)

8) Passes through
   (8, 1), (9, 4)

9) Passes through
   (-6, 4), (-2, 3)

10) Passes through
    (7, 4), (9, 7)

11) Slope = 5
    passes through (1, 2)

12) Slope = $-\frac{1}{2}$
    passes through (3, 4)

13) Passes through
    (3, 2), (-4, -5)

14) Passes through
    (-4, -3), (-8, -7)

REVIEW

Solve each equation.

1) $\frac{3x}{2} = 9$

2) $3\,(2x + 4) = -36$

3) $4x + 2 = 2x + 10$

4) $3x - 5 = -2$

Chapter 3 Review: Graphing and Analyzing Linear Equations

1) Draw a coordinate plane and graph the following points.

(4, 3), (5, 0), (7, -2), (-6, 4), (-1, 0), (-8, -6)

For questions 2 and 3, select 4 values for x, find the values for y, then plot the ordered pairs and connect them with a line.

2) $y = 2x + 3$

3) $y = \frac{x}{2} + 3$

For questions 4 and 5, find the x and y intercept for each line.

4) $4x + 6y = 12$

5) $y = -2x + 8$

For questions 6 and 7, find the slope of the line that passes though the points.

6) (4, 7), (3, 9)

7) (- 4, -5), (6, -3)

For questions 8 - 11, put each into an equation in the **slope-intercept form, y = mx + b**. Draw the graph.

8) $4y + 8x = -16$

9) slope 2, passes through (2, 5)

10) slope 2, passes through (3, -2)

11) Passes through (2, -3) (4, 0)

For questions 12 - 13, find the equation of a line in the **point-slope form**.

$$y - y_1 = m\,(x - x_1)$$

12) slope 3, passes through (1, -5)

13) Passes through (0, 2) (-1, 0)

Some Easy Ways to Stay Tuned-Up!

You are learning lots of algebra topics in this book. As you work through each chapter, you might find it helpful to go back occasionally and do some review on your own. Doing this can keep the topics that you have studied, fresh in your mind.

Here are a few tips that can help you to remember what you have learned.

- Watch the Online Video Tutorials a second time. Don't be passive. Work right along with the instructor.

- For each lesson that you have completed, go back and review the Introduction, Helpful Hints, and Examples.

- For chapters that you have completed, re-work the Chapter Review a second time. If a certain problem gives you difficulty, you can go back to the corresponding lesson to refresh your memory.

- Another helpful learning tool is the Glossary, which is located in the back of the book. The glossary contains important terms and definitions. It is a good place to review topics that you have learned, and also it can give you a brief introduction to terms that you will be studying later.

4-1 Meaning, Symbols, and Properties of Inequalities

INTRODUCTION

An **inequality** is a math sentence that states that one expression is greater than or less than another. Basically, inequalities are solved using the same rules that are used to solve equations. Remember the following information about inequalities.

> is read "is greater than"	5 > 1 is read "5 is greater than 1"
< is read "is less than"	1 < 5 is read "1 is less than 5"
≥ is read "is greater than or equal to"	5 ≥ 5 is read "5 is greater than or equal to 5"
≤ is read "is less than or equal to"	5 ≤ 5 is read "5 is less than or equal to 5"

Notice that the inequality sign always points to the smaller value, unless the two expressions are equal.

The further left a number is on the number line, the smaller it's value.

Inequalities can be easily graphed on a number line. Here are some examples of how we graph inequalities.

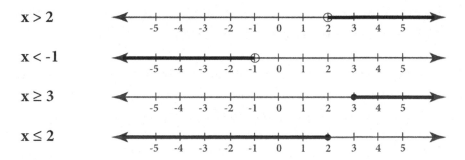

Notice that > and < show an **open** circle on the graph. The circled number is **not** part of the solution. Also notice that ≥ and ≤ show a **closed** circle on the graph. The number covered by the closed circle **is** part of the solution.

Remember the following tips when solving inequalities.

Helpful Hints
- Memorize the inequality signs.
- The inequality sign points to the smaller value.
- Graphs having < or > will have an open circle.
- Graphs having ≤ or ≥ will have a closed circle.
- Use graph paper when doing your work.

4-1 Meaning, Symbols, and Properties of Inequalities

EXAMPLES

Graph each of the following inequalities.

1) n < 4

2) n > -2.5

3) n ≤ -3.5

4) n ≥ 4

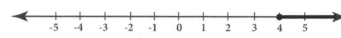

EXERCISES

Graph each of the following inequalities.

1) x > 7

2) x < -2

3) x ≥ 1

4) x ≤ -3

5) x > 1.5

6) x < -2.5

7) x ≤ -2.5

8) $x \geq 2\frac{1}{2}$

9) x > 4

10) x ≤ -2

11) x ≤ 0

12) $x > -3\frac{1}{2}$

13) x < 2.5

14) x ≤ -.5

15) $x \geq 3\frac{1}{4}$

16) x > -5.5

17) x ≤ -1.5

18) $x > 5\frac{3}{4}$

REVIEW

Change each equation to the slope-intercept form and sketch the graph.

1) y – x = 1

2) y + 2x = 8

3) 2y + 4x = -4

4) 4x + y = 5

4-2 Solving Inequalities Using Addition and Subtraction

INTRODUCTION

The rules for using addition and subtraction when solving inequalities are the same rules that are used when solving equations. Once the inequality has been solved it can be easily graphed on a number line. Remember the following.

Helpful Hints
- Adding or subtracting the same number on each side of the inequality sign will **not** change the direction of the inequality sign.
- Graphs showing > or < will have an **open** circle.
- Graphs showing ≥ or ≤ will have a **closed** circle.
- Use graph paper when doing your work.

EXAMPLES

Solve each inequality. Then graph the solution on a number line.

1) $x + 6 \geq -3$

    $x + 6 - 6 \geq -3 - 6$ *Subtract 6 from both sides*

       $x \geq -9$

2) $x - 3 < 2$

    $x - 3 + 3 < 2 + 3$ *Add 3 to both sides*

       $x < 5$

3) $x - -7 < -3$

      $x + 7 < -3$ *Rewrite*

    $x + 7 - 7 < -3 - 7$ *Subtract 7 from both sides*

       $x < -10$

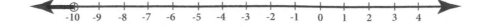

4-2 Solving Inequalities Using Addition and Subtraction

Solve each inequality. Then graph the solution on a number line.

1) $x + 7 < 10$

2) $x - 3 > -4$

3) $x - 7 \geq -5$

4) $x + 4 \leq 3$

5) $x - 3 < -5$

6) $x + 3 > -2$

7) $x - 3 \leq -5$

8) $x + 3 \geq 5$

9) $x - 12 \leq -6$

10) $x + 6 > -6$

11) $x - 1.5 < 3.5$

12) $x - \frac{1}{2} < 3$

13) $x - -8 < 7$

14) $x - -9 \geq -4$

15) $x + \frac{1}{2} > 3\frac{1}{2}$

16) $x + 1\frac{1}{2} \geq 2\frac{1}{2}$

17) $x - 9 > -7$

18) $x + 9 \leq -15$

REVIEW

Find the equation of the line in slope-intercept form $y = mx + b$

1) Slope = 5, passes through (5, 5)

2) Passes through (1, 1) and (3, 3)

4-3 Solving Inequalities Using Multiplication and Division

Be careful when using multiplication or division to solve an inequality. The following example will illustrate this.

$5 > 2$	*We know that 5 is greater than 2.*
$-2 \cdot 5 > -2 \cdot 2$	*Let's multiply both sides by -2.*
$-10 \not> -4$	*We know -10 is <u>not</u> greater than -4.*
$-10 < -4$	*So, we reverse the symbol*
	Keep this in mind when solving inequalities.

Remember the following tips.

Helpful Hints
- Multiplying or dividing each side of the inequality sign by **any positive number** will **not** change the direction of the inequality sign.

- Multiplying or dividing each side of the inequality sign by **any negative number** REVERSES the direction of the inequality sign.

EXAMPLES

Solve each inequality. Then graph the solution on a number line.

1) $5x \leq 30$

$$\frac{5x}{5} \leq \frac{30}{5} \quad \textit{Divide both sides by 5}$$

$$x \leq 6$$

2) $\dfrac{x}{-3} > 2$

$-3 \cdot \dfrac{x}{-3} > -3 \cdot 2$ *Multiply both sides by -3*

$x < -6$ *Since we multiplied by a negative number, reverse the symbol.*

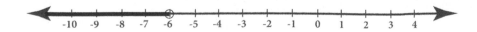

4-3 Solving Inequalities Using Multiplication and Division

Solve each inequality. Then graph the solution on a number line.

1) $5x \geq -25$

2) $-4x \leq -16$

3) $\frac{x}{2} \geq 20$

4) $\frac{x}{-2} < 14$

5) $2x > -8$

6) $-2x < 8$

7) $\frac{x}{-2} < 4$

8) $-7n < 21$

9) $\frac{n}{5} \leq -3$

10) $7n \geq 28$

11) $-10x > -20$

12) $\frac{1}{2}x \leq 8$

13) $.5x > 2.5$

14) $\frac{y}{-3} \leq -1$

15) $1.5x \geq 6$

16) $-1.5x < 3$

17) $\frac{x}{-2} \geq -4$

18) $\frac{x}{3} \leq -9$

REVIEW

Simplify each of the following. Leave answers as exponents.

1) $\frac{7^9}{7^3}$

2) $(7^2)^3$

3) $7^{11} \cdot 7^{13}$

4) $(n^3)^4$

INTRODUCTION

When solving **multi-step inequalities**, use the same techniques used to solve equations. It is recommended to keep the variable on the left side of the inequality sign when possible. Keep the following in mind.

Helpful Hints

- Adding or subtracting the same number on each side of the inequality sign will **not** change the direction of the inequality sign.

- Multiplying or dividing each side of the inequality sign by **any positive number** will **not** change the direction the inequality sign.

- Multiplying or dividing each side of the inequality sign by **any negative number REVERSES** the direction of the inequality sign.

- Use graph paper when doing your work.

EXAMPLES

Solve each inequality. Then graph the solution on a number line.

1) $2x + 5 > 3$

$2x + 5 - 5 > 3 - 5$ *Subtract 5 from both sides*

$2x > -2$

$\dfrac{2x}{2} > \dfrac{-2}{2}$ *Divide both sides by 2*

$x > -1$

2) $-3x + 9 \leq 3$

$-3x + 9 - 9 \leq 3 - 9$ *Subtract 9 from both sides*

$-3x \leq -6$

$\dfrac{-3x}{-3} \leq \dfrac{-6}{-3}$ *Divide both sides by -3*

$x \geq 2$ *Reverse inequality sign since we divided both sides by a negative number*

3) $8x + 20 \leq 3x - 25$

$8x - 3x + 20 \leq 3x - 3x - 25$ *Subtract 3x from both sides*

$5x + 20 \leq -25$

$5x + 20 - 20 \leq -25 - 20$ *Subtract 20 from both sides*

$5x \leq -45$

$\dfrac{5x}{5} \leq \dfrac{-45}{5}$ *Divide both sides by 5*

$x \leq -9$

4) $3(2x - 5) \geq 8x - 5$

$\overset{\frown}{3(2x - 5)} \geq 8x - 5$ *Use the distributive property to simplify the equation*

$6x - 15 \geq 8x - 5$

$6x - 8x - 15 \geq 8x - 8x - 5$ *Subtract 8x from both sides*

$-2x - 15 \geq -5$

$-2x - 15 + 15 \geq -5 + 15$ *Add 15 to both sides*

$-2x \geq 10$

$\dfrac{-2x}{-2} \geq \dfrac{10}{-2}$ *Divide both sides by -2*

$x \leq -5$ *Reverse inequality sign since we divided by a negative.*

Notice, we keep the variable on the left

EXERCISES

Solve each inequality. Then graph the solution on a number line.

1) $3x - 8 > -14$

2) $2x + 6 < 12$

3) $-4x - 8 \leq 10$

4) $\dfrac{x}{3} + 2 > 3$

5) $\dfrac{m}{-3} - 6 \leq 1$

6) $-5m + 3 \geq 28$

7) $4x - 1 \geq x + 8$

8) $3x - 4 < 6x + 2$

9) $4(2x - 3) \leq -3x - 1$

10) $5x + 7 > 4x - 5$

REVIEW

Simplify each of the following.

1) $\dfrac{1}{5} - \dfrac{4}{5}$

2) $-.7 - .9$

3) $5 \bullet -3.4$

4) $\dfrac{-.15}{-.3}$

4-5 Solving Combined Inequalities

Combined inequalities are joined by the words **and** or **or**.

A **conjunction inequality** is joined by the word **and**. In this type of inequality there are two parts. It is necessary to solve both parts separately. A conjunction type of inequality is true when **both** of the inequalities are true. The inequality $-4 < x < 5$ would be read **"x is greater than -4 and x is less than 5."** Notice that the inequality signs point to the smaller values.

A disjunction inequality is joined by the word **or**. In this type of inequality it is also necessary to solve both parts separately. A disjunction type of inequality is true when **at least one** of the inequalies is true.

Remember the following when solving combined inequalities.

Helpful
Hints
- Solve each part of the inequality separately.
- Graph each solution on a number line.
- The inequality sign points to the smaller value. Remembering this can help when reading inequalities.
- Practice reading conjunction types of inequalities. For example, $3 < x + 1 < 6$ would be read **"x + 1 is greater than 3 and x + 1 is less than 6."**

EXAMPLES

Solve each combined inequality. Then draw the graph on a number line.

1) $-2 < x + 4 < 7$

$-2 < x + 4$ *Subtract 4 from both sides.*   and $x + 4 < 7$

$-2 - 4 < x + 4 - 4$ $x + 4 - 4 < 7 - 4$ *Subtract 4 from both*

$-6 < x$ and $x < 3$

The solution is $-6 < x < 3$.

This is read "x is greater than -6 and x is less than 3."

Graph the inequality.

4-5 Solving Combined Inequalities

2) $2x - 3 \leq -11$ or $2x - 12 > 0$

$2x - 3 \leq -11$	or	$2x - 12 > 0$
$2x - 3 + 3 \leq -11 + 3$ *Add 3 to both sides.*		$2x - 12 + 12 > 0 + 12$ *Add 12 to both sides*
$2x \leq -8$		$2x > 12$
$\dfrac{2x}{2} \leq \dfrac{-8}{2}$ *Divide both sides by 2.*		$\dfrac{2x}{2} > \dfrac{12}{2}$ *Divide both sides by 2.*
$x \leq -4$	or	$x > 6$

Graph each inequality.

EXERCISES

Solve each combined inequality. Then draw the graph on a number line.

1) $-2 < x + 1 < 3$

2) $-7 \leq x + 5 < 2$

3) $x + 1 > 2$ or $x + 1 < -6$

4) $-2 < 2x + 4 \leq 8$

5) $-5 \leq 3x + 1 \leq 4$

6) $x + 5 \leq -2$ or $x + 5 \geq 2$

7) $-2 < x + 3 < 7$

8) $4x + 2 > 10$ or $2x < -10$

9) $-6 \leq -2x + 2 < 8$

10) $2x + 1 \leq -3$ or $2x + 1 > 3$

REVIEW

Simplify. Leave the answer as an exponent.

1) $x^3 \cdot x^5$

2) $9n^3 \cdot 3n^5$

3) $(x^4)^3$

4) $\dfrac{6^9}{6^2}$

4-6 Solving Inequalities Involving Absolute Value

From our work with equations we know that the **absolute value** of a number is its distance from 0 on a number line. The expression $|-3|$ is read "the absolute value of -3." The absolute value of a number is never negative.

Remember the following when solving and graphing inequalities involving absolute value.

Helpful Hints
- Solving inequalities involving absolute value is very similar to solving equations involving absolute value.
- Inequalities involving absolute value will have two solutions.
- Use graph paper when doing your work.

EXAMPLES

Solve each of the following inequalities. Then graph the solution on a number line.

1) $|x| > 5$

$x < -5 \quad \underline{or} \quad x > 5$

To test your answer, you could take any number between -5 and 5. Its absolute value would <u>not</u> be greater than 5.

2) $|x| < 5$

$x > -5 \quad \underline{and} \quad x < 5$

Rewrite as $-5 < x < 5$

To test your answer, take any number to the left of -5 or to the right of 5. Its absolute value would not be less than 5.

3) $5|x| - 22 > 13$

$5|x| - 22 + 22 > 13 + 22$ *Add 22 to both sides*

$5|x| > 35$

$\dfrac{5|x|}{5} > \dfrac{35}{5}$ *Divide both sides by 5*

$|x| > 7$

$x < -7 \quad \underline{or} \quad x > 7$

Test your answer by substituting a number to the left of -7 and to the right of 7.

4) $|x + 1| \leq 3$

$x + 1 \geq -3 \qquad \underline{and} \qquad x + 1 \leq 3$

$x + 1 - 1 \geq -3 - 1$ *Subtract 1 from both sides* $\quad x + 1 - 1 \leq 3 - 1$ *Subtract 1 from both sides*

$x \geq -4 \qquad \underline{and} \qquad x \leq 2$

Rewrite as $-4 \leq x \leq 2$

To test your answer, substitute any number to the right of -4 or to the left of 2.

4-6 Solving Inequalities Involving Absolute Value

Solve each of the following inequalities. Then graph the solution on a number line.

1) $|x| > 10$

2) $|x| \leq 10$

3) $2|x| < 12$

4) $|x + 4| > 6$

5) $|x - 6| < 10$

6) $2|x - 3| \geq 6$

7) $|x - 6| \leq 6$

8) $|5x| > 15$

9) $4|x| - 1 < 15$

10) $3|x| - 2 \geq 7$

11) $|x - 11| < 4$

12) $|2n - 5| \leq 11$

13) $|y + 9| \geq 7$

14) $|4x - 9| < 3$

15) $|-3 + n| < 18$

16) $|15x - 15| - 4 \geq 21$

17) $|3n - 9| - 2 \leq 7$

18) $|4n + 2| < 6$

REVIEW

Simplify each of the following.

1) $-\frac{3}{4} + \frac{1}{2}$

2) $-\frac{1}{2} - \frac{1}{4}$

3) $\frac{1}{2} \cdot -1\frac{1}{2}$

4) $-3\frac{1}{2} \div \frac{1}{2}$

4-7 Graphing Linear Inequalities

INTRODUCTION

Graphing linear inequalities is similar to graphing linear equations. First draw the graph of the line. The graph of the line will divide the **coordinate plane** into two **half planes**. If the inequality contains ≤ or ≥, the line will be **solid**, meaning **it is part** of the solution. If the inequality contains < or >, the line will be **dashed**, meaning **it is not part** of the solution.

To determine which side of the line to shade, it is easy to substitute a point into the inequality. Use the following steps when graphing inequalities.

Helpful Hints
- **First**, make sure that y is on the left of the inequality sign. That way you will know the y-intercept and the slope to make it easy to graph the line.

- **Second**, draw a solid line if ≤ or ≥ are used. Draw a dashed line if < or > is used.

- **Third**, shade the appropriate region. Often it is easy to substitute the origin (0, 0) into the inequality.

EXAMPLES

Graph each inequality.

1) **y < 3x + 1**

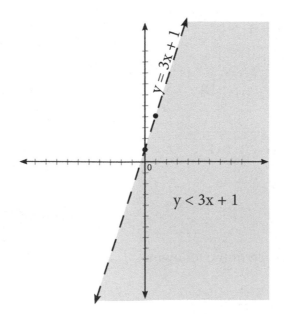

The slope is 3 and the y intercept is 1.

Since < is used, the line is dashed. It is not part of the solution. Draw the line.

Substitute (0, 0) into the inequality to see which side to shade.

$$y < 3x + 1$$

$$0 < 3(0) + 1$$

$$0 < 1$$

This is true, which means (0, 0) is part of the shaded region.

Shade the region below the line.

2) -2x + y ≥ 4

Solve for y.

$y \geq 2x + 4$

The slope is 2 and the y intercept is 4.

Since ≥ is used, the line is solid.
It is part of the solution.

Substitute (0, 0) into the inequality.

$y \geq 2x + 4$

$0 \geq 2(0) + 4$

$0 \geq 4$

This is not true, which means
(0, 0) is not part of the solution.

Shade the region above the line.

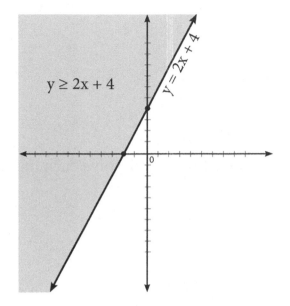

EXERCISES

Graph each linear inequality.

1) $y < 3x + 2$ 2) $y > 2x - 2$

3) $y \leq 4x + 2$ 4) $y \geq 2x - 4$

5) $y - 2x < 1$ 6) $2y - 2x \leq 6$

7) $y \leq x + 5$ 8) $y \geq 3x - 2$

9) $y - 3x \leq -2$ 10) $y - x \geq 4$

REVIEW

Draw the graph of each inequality on a number line.

1) $x > 4$ 2) $x \leq -2$

3) $2 < x < 4$ 4) $x < 2$ or $x \geq 5$

Chapter 4 Review: Solving and Graphing Inequalities

For questions 1–4, sketch the graph of each on a number line.

1) $x \leq 4$

2) $x \geq -3$

3) $x > -5$

4) $x < 1$

For questions 5–15, solve each inequality and graph each solution on a number line.

5) $x + 3 > 2$

6) $x - 7 \leq -2$

7) $3x \geq -12$

8) $-2x < 10$

9) $-8x - 16 \leq 20$

10) $\frac{x}{-6} - 12 \leq 2$

11) $|x + 5| \geq 8$

12) $|x| > 8$

13) $|x| \leq 8$

14) $4|x + 2| > 16$

Graph each of the following inequalities on a coordinate plane.

15) $y > 3x - 2$

16) $y \leq 3x + 2$

One of the Most Important Tips of All!

This tip applies to this book, and to any other math book
that you use in the future.

Here it is.

Most math books contain examples that show the necessary steps in solving
a problem. Many students merely read through the examples. Then, when
they have finished reading and attempt to do the written exercises, they
experience difficulties, and wonder why.

This is what you need to do. First, neatly copy the example on a sheet of
paper. Next, also copy each and every step.

I promise that by doing this, you will find it much easier to understand the
problem, and you will remember the process necessary to solve it.

There is something very special about writing a problem down and then
writing out the steps. It makes the learning process so much more effective.
When you do this, you are fully involved and will experience a much deeper
understanding.

Just simply reading a problem and the steps is not nearly as effective.

This is a Very Simple Tip. But It Works!

5-1 The Graphing Method

INTRODUCTION

Two or more linear equations is a called a **system of linear equations**. There are several ways to solve a system of linear equations. **Graphing** is one of the ways. Remember that the graph of any linear equation is a line. The solution set of a pair of linear equations is the **point** on the coordinate system where the two lines intersect. That means that the **solution** will be an **ordered pair**. To solve a system of linear equations using the graphing method, follow these steps.

Helpful Hints
- **First**, graph one equation on a coordinate plane.
- **Second**, graph the other equation.
- The point of intersection of the two lines is the solution set.
- Check by substituting the ordered pair into each equation.
- It is very helpful to use graph paper. Your graph will be much more accurate.

EXAMPLE

Use the graphing method to find the solution for $2x + y = 8$ and $y - x = 2$.
First select a few values for x and then find y for each equation.
Then draw the graph of each.

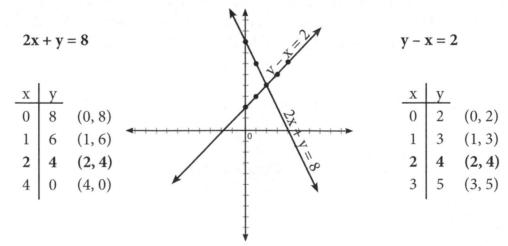

$2x + y = 8$

x	y	
0	8	(0, 8)
1	6	(1, 6)
2	**4**	**(2, 4)**
4	0	(4, 0)

$y - x = 2$

x	y	
0	2	(0, 2)
1	3	(1, 3)
2	**4**	**(2, 4)**
3	5	(3, 5)

The point of intersetion is the ordered pair (2, 4).

Now check by substituting the ordered pair into each equation.

$2x + y = 8$
$2(2) + 4 = 8$
$4 + 4 = 8$
$8 = 8$

Both are true.

$y - x = 2$
$4 - 2 = 2$
$2 = 2$

The solution is (2, 4).

5-1 The Graphing Method

Use the graphing method to find the ordered pair that is the solution to each system of linear equations. For an accurate graph, use graph paper.

1) $y = x + 1$
   $y = -x + 3$

2) $y = 2x$
   $y = -2x - 4$

3) $y = x + 4$
   $y = 2x + 5$

4) $x + y = 7$
   $x - y = 1$

5) $y - x = 3$
   $x + y = -1$

6) $x + y = 4$
   $x - y = 0$

7) $y = 2x - 5$
   $y = 7 - x$

8) $y = -x + 8$
   $y = 4x - 7$

9) $x + y = 4$
   $2x + y = 5$

10) $2x + y = 8$
    $y - x = 2$

11) $y = 2x + 5$
    $y = x + 4$

12) $y = 3x - 3$
    $y = 2x$

13) $x - 2y = 4$
    $y - x = -2$

14) $x - y = 6$
    $x + y = -4$

REVIEW

Solve each inequality and draw the graph of the solution on a number line.

1) $2x - 4 > 12$

2) $-2x - 4 \leq 5$

3) $\frac{m}{2} + 3 \geq 10$

4) $\frac{x}{-3} + 4 < 1$

5-2 The Substitution Method

Another method of solving systems of linear equations is the **substitution method**. This method is especially good to use when the coefficient of one of the variables is 1 or -1.

Remember the following steps when using the substitution method.

Helpful Hints
- **First**, solve one of the equations for either x or y.
- **Second**, substitute the expression in the other equation and solve it. This will give you the value of one of the variables.
- **Third**, take this value and substitute it into either one of the original equations to get the value of the second variable.
- **Fourth**, check the values for x and y by substituting them into the original equations.
- **Remember**, the solution will be an ordered pair.

EXAMPLES

Use the substitution method to solve each system of linear equations.

1) $4x + 3y = 27$
         $y = 2x - 1$

Substitute $(2x - 1)$ for y in the first equation and solve for x.

$4x + 3(2x - 1) = 27$ *Use the distributive property*

$4x + 6x - 3 = 27$ *Collect like items*

$10x - 3 = 27$ *Add 3 to both sides*

$10x = 30$ *Divide both sides by 10*

$x = 3$

Now substitute x = 3 into either equation to get the value of y.
Let's use the second equation.

$y = 2x - 1$

$y = 2(3) - 1$

$y = 6 - 1$

$y = 5$ So, x = 3, y = 5 = (3, 5)

The solution is the ordered pair (3, 5)

Check by substituting these values into the original equations.

5-2 The Substitution Method

2)
$$5x + 3y = 17$$
$$x - 2y = 6 \longleftarrow \textit{First rewrite this as } x = 2y + 6$$

Substitute $(2y + 6)$ for x in the first equation and solve for y.

$5(2y + 6) + 3y = 17$ *Use the distributive property*

$10y + 30 + 3y = 17$ *Collect like items*

$13y + 30 = 17$ *Subtract 30 from both sides*

$13y = -13$ *Divide both sides by 13*

$y = -1$

Now substitute y = -1 into the second equation (since it's simpler) and solve for x.

$x = 2y + 6$

$x = 2(-1) + 6$

$x = -2 + 6$

$x = 4$ So, x = 4, y = -1 = (4, -1)

The solution is the ordered pair (4, -1)

Check by substituting these values into the original equations.

EXERCISES

Use the substitution method to solve each system of linear equations.

1) $x = y + 3$
   $x + 7 = 2y$

2) $y = 2x$
   $3x + y = 10$

3) $y = 3x$
   $5x - 2y = 1$

4) $y = x + 4$
   $3x + y = 16$

5) $x + y = 2$
   $3x + y = 8$

6) $y - 3x = 1$
   $4x + y = 8$

7) $x + y = 5$
   $2x + y = 6$

8) $x = 2y$
   $3x = y - 10$

9) $x = 3y$
   $3y + 2x = 18$

10) $x + 3y = 0$
    $2x + 9y = 10$

11) $x = y + 4$
    $2x - 5y = 8$

12) $y = 3x + 1$
    $2x + 3y = 25$

REVIEW

Solve each inequality. Then draw a graph on a number line.

1) $|x + 2| > 8$

2) $|x - 3| \leq 6$

3) $|x| < 8$

4) $|x| > 3$

5-3 The Elimination Method

INTRODUCTION

Another method for solving systems of linear equations is the **elimination method**. Sometimes it is possible to add or subtract the equations to get a new equation with only one variable. Sometimes this method is called the **addition/subtraction method**. If two equations have the same or opposite coefficients for one of the terms, the elimination method can be used. Remember the following steps when using this method.

Helpful Hints
- **First**, make sure the variables are on one side and the constant on the other, with the like terms lined up.

- **Second**, add or subtract the equations to eliminate one of the variables.

- **Third**, solve the equation.

- **Fourth**, substitute the answer into either of the equations to get the value of the second variable.

- Check by substituting the answers into the original equations.

- Sometimes it is necessary to multiply one of the equations by a constant. You will see this in one of the examples.

EXAMPLES

Solve each of the following using the elimination method.

1) $x + 2y = 5$
   $-x + y = 13$

$$\begin{array}{r} x + 2y = 5 \\ + \underline{-x + y = 13} \\ 3y = 18 \\ y = 6 \end{array}$$
   First, add the equations

   Solve for y

$$\begin{array}{r} x + 2y = 5 \\ x + 2(6) = 5 \\ x + 12 = 5 \\ x = -7 \end{array}$$
   Substitute y = 6 into either equation. Let's use the first equation.

   So, $x = -7$, $y = 6$ = $(-7, 6)$

The solution is the ordered pair $(-7, 6)$

Check by sustituting your answer into the original equations.

5-3 The Elimination Method

2) $5x + y = 13$
       $4x - 3y = 18$

First, multiply the first equation by 3. Remember to mulitply both sides.
$3(5x + y) = 3 \cdot 13 \quad = \quad 15x + 3y = 39$

$\begin{array}{r} 15x + 3y = 39 \\ + \quad 4x - 3y = 18 \\ \hline 19x = 57 \\ x = 3 \end{array}$ *Now add the equations.*

Solve for x.

Next, substitute x = 3 into either equation.
Let's use the first equation and solve for y.

$5x + y = 13$
$5(3) + y = 13$
$15 + y = 13$
$y = -2$ So, x = 3, y = -2 = (3, -2)

The solution is the ordered pair (3, -2).
Check by sustituting your answer into the original equations.

EXERCISES

Solve each of the following using the elimination method.

1) $x + 6y = 10$
   $x + 2y = 2$

2) $3x + y = 7$
   $-2x + y = -8$

3) $2x + y = 10$
   $3x - y = 5$

4) $4x - 7y = 13$
   $4x + 7y = -29$

5) $5x + 3y = 14$
   $2x + y = 6$

6) $3x + y = 6$
   $x + 3y = 10$

7) $-x + y = 3$
   $3x + 2y = 26$

8) $3x - y = 1$
   $x + 2y = 12$

9) $2x - y = -6$
   $-x + y = 2$

10) $2y + 5x = 10$
    $-y + x = 2$

11) $x + 2y = 4$
    $3x + y = 7$

12) $3x + y = 6$
    $x + 3y = 10$

13) $2x + y = 12$
    $x + 2y = 9$

14) $x + y = 5$
    $2x + y = 6$

15) $x - 3y = -11$
    $3x + y = 17$

REVIEW

Select a few values for x and find the values for y. Then graph the linear inequality on a coordinate plane. (Hint: You are given the slope and y intercept. Also, test your solution using the origin, (0, 0).)

1) $y < x + 1$

2) $y > 2x + 3$

5-4 Graphing Systems of Inequalities

INTRODUCTION

Two or more linear inequalities are a **system of inequalities**. To find the solution set of a pair of linear inequalities, we must find the ordered pairs that satisfy both of the inequalities. To do this, graph and appropriately shade each inequality. The region where the two shaded areas overlap is the solution set. Use the following steps when graphing a system of linear inequalities. It might be helpful to review lesson **4-7 Graphing Inequalities**.

Helpful Hints
- **First**, make sure that y is on the left of each inequality sign. That way you will know the y-intercept and the slope. Sometimes you may want to find the x and y-intercepts and then graph each inequality.
- **Second**, graph the first inequality and shade the appropriate region.
- **Third**, graph the second inequality and shade the appropriate region.
- When drawing the graphs remembers to draw a solid line if ≤ and ≥ are used. Draw a dashed line if < and > are used.
- The region where the two shaded areas overlap is the solution set.
- Check your work by using a point from the overlapped area. Substitute it into each inequality to make sure you shaded correctly.
- Use graph paper when doing your work.

EXAMPLES

Graph each pair of inequalities. The solution set will be where the two shaded regions overlap. Use the slope-intercept form or find the x and y-intercepts to help in drawing the graphs.

1) $y \leq x + 1$
   $y \geq 2 - x$

First, graph each inequality and shade appropriately.

Let's check our solution by taking a point (3, 0) from the overlap area and substituting it into each inequality.

$y \leq x + 1$	$y \geq 2 - x$
$0 \leq 3 + 1$	$0 \geq 2 - 3$
$0 \leq 4$	$0 \geq -1$

Both are true. The graph is correct.

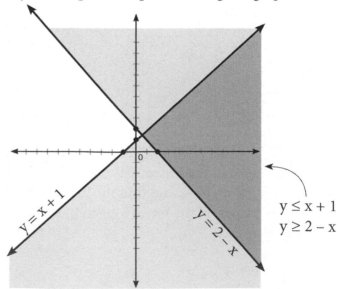

$y = x + 1$

$y = 2 - x$

$y \leq x + 1$
$y \geq 2 - x$

2) $y < 3x + 1$

$y < -x - 4$

First, graph and shade.

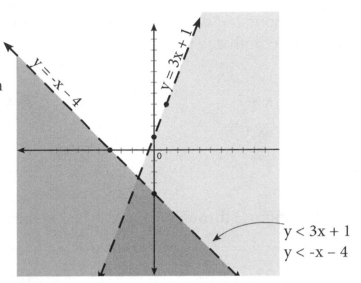

Let's test by taking point (0, -6) from the overlap area and substituting it into each inequality.

$y < 3x + 1$	$y < -x - 4$
$-6 < 3(0) + 1$	$-6 < 0 - 4$
$-6 < 1$	$-6 < -4$

Both are true. The graph is correct.

$y < 3x + 1$
$y < -x - 4$

EXERCISES

Graph each pair of linear inequalities. The solution set of the system is where the shading overlaps.

1) $y > -x + 8$
   $y \geq 4x - 7$

2) $y < x + 2$
   $y > -2x - 1$

3) $y < 2x + 1$
   $y \leq 4x + 1$

4) $y < 3$
   $y \geq x + 1$

5) $y > 2x + 2$
   $y \geq -x - 1$

6) $x + y \geq 4$
   $y \leq 2x - 3$

7) $x + y \geq 2$
   $y \geq x - 1$

8) $y \geq -x + 4$
   $y \leq 2x - 3$

REVIEW

Solve each of the following using the substitution method.

1) $y = 4x$
   $x + y = 10$

2) $y = x - 2$
   $x + y = 12$

Chapter 5 Review: Systems of Linear Equations

For questions 1 and 2, use the graphing method to find the ordered pair that is the solution to both equations.

1) $y = x + 1$
   $y = -x + 3$

2) $y = 2x + 6$
   $y = -x - 3$

For questions 3 through 6, solve by using the substitution method.

3) $y = 2x$
   $x + y = 9$

4) $y = x + 1$
   $x + y = 5$

5) $y = 3x + 1$
   $4x + y = 8$

6. $x + y = 2$
   $3x + y = 8$

For questions 7–10, solve by using the elimination method.

7) $4x - y = 8$
   $2x + y = -2$

8) $2x + 3y = 7$
   $2x - y = -5$

9) $2y + 5x = 10$
   $-y + x = 2$

10) $2x - y = 5$
    $x - 2y = 1$

For questions 11–12, graph each system of linear inequalities.

11) $x + y > 2$
    $y < x + 3$

12) $y \leq x + 4$
    $y < -2x + 6$

What About the Problems that I Get Wrong?

When learning algebra, every student makes mistakes.

It is important to find out why you worked a problem incorrectly, so that you don't make the same mistake in the future.

Here are some tips to help you correct your mistakes.

■ Many mistakes are based on careless errors. Carefully re-working the problem can usually result in finding these errors.

Here are just a few common careless errors:

- The student did not carefully read the problem and understand what was asked.

- The problem was not copied or set up correctly.

- There were mistakes with positive or negative signs.

- There were mistakes in the order of operations.

- Fractions were not reduced to lowest terms.

- Answers were not in their simplest form.

- Careless computational errors were made.

■ Sometimes the student does not fully understand the topic. This will result in difficulties with the problems. If this is the case, it would be good to consider doing any or all of the following:

- View and the Online Tutorial Video again, being careful to work along with the instructor.

- Re-read the Introduction and Helpful Hints section.

- Study the Examples pertaining to the topic, being careful to copy the problem as well as all the steps.

6-1 Adding Polynomials

INTRODUCTION

A **polynomial** is an algebraic expression of one or more **terms** connected by plus (+) and minus (–) signs. A **monomial** has **one term**. A **binomial** has **two terms**. A **trinomial** has **three terms**. Here are some examples.

monomials	$3xy$	$-3x$	$3x^2y$
binomials	$4a + 2b$	$2y - 3x\,y$	$3x + 7xyz$
trinomials	$5x - 12xy + y$	$3x^3 + 6x - 7$	

The **degree** of a polynomial is the highest of the degrees of its terms. In the example $3x^3 + 2x^2 - xy$, the degree is 3. Notice that polynomials are often written in **descending order**. The highest degree is on the left and the lowest is on the right.

Like terms have the same variables with the same degrees. For example $3x^2y$ and $12x^2y$ are like terms. The coefficients **can** be different.

To add polynomials, simply combine like terms by adding their coefficients. Remember the following when adding polynomials.

Helpful Hints
- To add polynomials in a vertical form, simply line up the like terms and add the coefficients. If there are any **unlike** terms left over, write them down.

- To add polynomials in a horizontal form, remove the parentheses and simply group the like terms and add the coefficients. If there are any **unlike** terms left over, write them down.

EXAMPLES

1) **Add using the vertical method.**

   $-2x^2 + 7x - 2y - 8$ **and** $4x^2 - 6x + 4$

$$
\begin{array}{r}
-2x^2 + 7x - 2y - 8 \\
4x^2 - 6x + \ 0 + 4 \\
\hline
2x^2 + \ x - 2y - 4
\end{array}
$$

2) **Add using the horizontal method.**

   $(8x^2y + 5x + 2y - 2z) + (3x^2y - 7x + 3y)$

   $8x^2y + 3x^2y + 5x - 7x + 2y + 3y - 2z$ *Remove the parentheses*

   $11x^2y - 2x + 5y - 2z$ *Collect the terms and add the coefficients*

6-1 Adding Polynomials

Add each of the following.

1) $10x^2 + 12xy + 4x$
$+ \ 12x^2 - 22xy + 8x$

2) $2xy + 5x + 6y$
$+ \ 4xy - 3x - 8y$

3) $(2x + 9) + (4x - 18)$

4) $(5y - 6) + (-7y + 12)$

5) $-9x^2 + 7x + 2$
$+ \ -2x^2 + 6x + 9$

6) $-6y^2 - 3y + 3x - 7$
$+ \ 5y^2 + 4y - 5x + 2$

7) $(4x^2 - 8x - 7) + (7x^2 - 7x - 9)$

8) $(-x^3 + 5x^2 + 6x + 8) + (4x^3 - 6x + 2)$

9) $(x^3 + 7x - 5) + (-3x^2 - 12x + 5)$

10) $3x^2y - 2xy + 3x + 4y$
$+ \ -2x^2y - 3xy - 5x + 7y$

11) $(x^2 + 7x - 14) + (-3x^2 - 5x + 6)$

12) $(3x^3 - 3x + 7) + (4x^3 + 2x^2 - 3x)$

13) $(-x^2 + 12x - 8) + (-3x^2 - 15x + 12)$

14) $7x^2 + 9x - 13$
$+ \ 4x^2 - 2x - 11$

REVIEW

Solve using the elimination method.

1) $y = 3x$
$y - x = 18$

2) $y = 3x - 1$
$7x + 2y = 37$

6-2 Subtracting Polynomials

Subtracting polynomials can be done either vertically or horizontally. Remember the following when subtracting polynomials.

Helpful Hints

- Subtracting **vertically** involves two steps.
  First, change the sign of each term that is being subtracted.
  Second, add them.

- When subtracting **horizontally** use the following steps.
  First, remove the parentheses and the minus sign (-).
  Second, change the sign of each term in the group being subtracted.
  Third, collect the like terms and add the coefficients.

EXAMPLES

Subtract each of the following.

1) Subtract $(8x^2 + 3x + 7) - (4x^2 - x + 1)$ vertically.

$$\begin{array}{r} 8x^2 + 3x + 7 \\ + \ -4x^2 + \ \ x - 1 \\ \hline 4x^2 + 4x + 6 \end{array}$$ ←*Change the sign of each term and add.*

2) Subtract $(5x^2 + 4x - 6) - (2x^2 - 2x - 2)$ horizontally.

First, remove the parentheses and the minus sign (–).

Second, change the sign of each term in the second group.

$= \ 5x^2 + 4x - 6 - 2x^2 + 2x + 2$

Now, collect like terms and add.

$= \ 5x^2 + -2x^2 + 4x + 2x - 6 + 2$

$= \ 3x^2 + 6x - 4$

6-2 Subtracting Polynomials

Subtract 1-10 vertically

1) $(2x + 9) - (4x - 18)$

2) $(9x^2 - 4x + 8) - (3x^2 - 5x - 7)$

3) $(x^2 + 5xy - 2z) - (3x^2 + xy - 4z)$

4) $(3n - 2) - (-n - 2)$

5) $(4x^2 - 3x + 5) - (x^2 - 3x + 2)$

6) $(5x - 3y) - (7x - 4y)$

7) $(3x - 7y) - (-5x + 3y)$

8) $(3x^2 - 3x + 6) - (x^2 + 4x - 10)$

9) $(7x - 8) - (9x + 4)$

10) $(7x^2 - 3x + 4) - (4x^2 - 5x - 6)$

Subtract 11-20 horizontally

11) $(2x - 4) - (x - 3)$

12) $(3x^2 + 5x - y^2) - (x^2 - 4x + 2y^2)$

13) $(x^2 - 7x + 6) - (5x^2 + 7)$

14) $(12x - 50) - (9x + 25)$

15) $(5x^2 - 12x - 3) - (-2x^2 + 10x + 10)$

16) $(x^2 - 4x + 5) - (2x^2 - 3x - 4)$

17) $(x^2 - 5x + 7) - (-3x^2 - 5x)$

18) $(3x^2 - 4x - 8) - (4x^2 + 7x + 3)$

19) $(7x^2 + 3x - 7) - (5x + 12)$

20) $(2x^2 - 3xy + 6) - (4x^2 + xy - 3)$

REVIEW

Find the slope of the line that passes through the given points.

1) $(2, 1), (3, 2)$

2) $(-1, -2), (4, -3)$

3) $(0, 5), (-4, -5)$

4) $(3, 4), (2, 2)$

6-3 Multiplying Monomials

A **monomial** is an algebraic expression that has one term. For example, 5, x, -3y, and 7xy are all monomials. To multiply monomials, **multiply the coefficients** together and **multiply the variables** together. Remember the following when multiplying monomials.

Helpful • To multiply powers of the same base, add the exponents.
Hints • To find the power of a power, multiply the exponents.

Multiply each of the following.

1) $5a^2 \cdot 3a^3$

$= 15a^{2+3}$

$= 15a^5$

2) $-5y^4 \cdot 6y^3$

$= -30y^{4+3}$

$= -30y^7$

3) $(xy)(x^2y^2)(2x^3y)$

Hint: $x = x^1$

$= 2x^{1+2+3}y^{1+2+1}$

$= 2x^6y^4$

4) $(3x^2)^3$

$= 3^3 \cdot x^{2 \cdot 3}$

$= 27x^6$

5) $(-4a^3)(-5a^5)$

$= -4 \cdot -5 \cdot a^{3+5}$

$= 20a^8$

6) $(r^2s^3t^4)(r^5st^3)$

$= r^{2+5} \cdot s^{3+1} \cdot t^{4+3}$

$= r^7s^4t^7$

6-3 Multiplying Monomials

Multiply each of the following.

1) $a^3 \cdot a^4$

2) $x^2 \cdot 3x^5$

3) $(3x)(-2x^4)$

4) $(xy)(3y^3)$

5) $(6x^2y^3)(-4x^4y^2)$

6) $(2r^2s^3)(3r^3s^2)(2r^5s^5)$

7) $(2a^2)^3$

8) $(-4a)^2$

9) $(m^3n)(-5m^3n^2p^5)$

10) $(3a^2)^3$

11) $(3x^2y^2)(4x^3y)$

12) $(2x)^2(3y)^2$

13) $(9x)^2$

14) $(-2y)(5xy)(4xy^2)$

15) $(-3x^4y^2)(7x^2y^5)$

16) $(-2y^2)(4y^4)(-3y^3)$

17) $(-2x^2y)^3$

18) $(2x)^2(3y^2)^2$

19) $(\frac{1}{2}x^3)(12x^7)$

20) $(\frac{3}{4}x^3)(12x^5)$

Add each of the following.

1) $(5x^4 + x^3 - 2x^2 + 7x - 5) + (-2x^3 + x^2 - 5x + 3)$

2) $(m^4 + m^3 - 2m^2 + 7m - 5) + (-3m^4 - m^3 - 2m^2 - 2m)$

6-4 Dividing Monomials

Dividing monomials is a simple process. The process is similar to simplifying or reducing fractions to their lowest terms, just like you learned in elementary school.

Remember the following when dividing monomials.

Helpful Hints

- **First**, divide the coefficients.

- **Second**, divide the variables.

- **Third**, simplify completely.

- Remember that any variable without an exponent means the same as the variable to the first power. Example: $x = x^1$

- $\dfrac{a^m}{a^n} = a^{m-n}$ 　　　　Example: $\dfrac{x^5}{x^2} = x^{5-2} = x^3$

- $a^{-n} = \dfrac{1}{a^n}$ 　　　　Example: $x^{-5} = \dfrac{1}{x^5}$

- You will often express your answer with positive exponents.

- Review the laws of exponents if necessary.

EXAMPLES

Divide each of the following. Express your answer with positive exponents.

1) $\dfrac{10x^3 y^3}{2x^2 y}$

$= \dfrac{10}{2} \bullet x^{3-2} \bullet y^{3-1}$

$= 5xy^2$

2) $\dfrac{-12a^2 b^5}{3a^4 b^2}$

$= -4 \bullet a^{2-4} \bullet b^{5-2}$

$= -4a^{-2} b^3$

$= \dfrac{-4b^3}{a^2}$

3) $\dfrac{(a^4 b^7)^4}{a^8 b^7}$

$= \dfrac{a^{16} b^{28}}{a^8 b^7}$

$= a^{16-8} \bullet b^{28-7}$

$= a^8 b^{21}$

4) $\dfrac{20a^3 c^4 d^2}{-5a^3 c^3}$

$= -4a^{3-3} \bullet c^{4-3} \bullet d^2$

$= -4 \bullet 1 \bullet c \bullet d^2$

$= -4cd^2$

6-4 Dividing Monomials

Divide each of the following. Express your answer with positive exponents.

1) $\dfrac{6a^3}{a^2}$

2) $\dfrac{-8a^3}{-2a}$

3) $\dfrac{24x^2y^2}{8xy}$

4) $\dfrac{36x^2y^4}{9x^4y^2}$

5) $\dfrac{(x^3y^3)^2}{x^4y^2}$

6) $\dfrac{-12a^3b^2}{3a^2b^3}$

7) $\dfrac{-12x^2y}{-2xy}$

8) $\dfrac{36a^3bc^4}{4a^2bc^6}$

9) $\dfrac{x^2y^3z^2}{y^4z^5}$

10) $\dfrac{25x^7y^6z^2}{5x^2yz}$

11) $\dfrac{24x^4y^2}{6x^3y^4}$

12) $\dfrac{-21x^5y^2z^3}{7x^4y}$

13) $\dfrac{20x^2y^3z}{-4x^4yz^3}$

14) $\dfrac{27x^2y^2}{3x^3y^3}$

15) $\dfrac{xy^2}{x^3y}$

16) $\dfrac{-75x^{12}y^7}{25x^4y^3}$

17) $\dfrac{24m^2n^3}{8mn^5}$

18) $\dfrac{15m^2n^2}{n}$

19) $\dfrac{25x^2y^3z^2}{5x^2yz^3}$

20) $\dfrac{9x^{12}y^{10}}{3x^7y^8}$

REVIEW

Subtract each of the following.

1) $(5x^4 + x^3 - 2x^2 + 7x - 5) - (-2x^3 + x^2 - 5x + 3)$

2) $(2y^5 + y^3 + 7y + 33) - (4y^5 - 4y^3 - 7y - 5)$

6-5 Multiplying Polynomials by Monomials

To multiply a polynomial by a monomial it is necessary to use the **distributive property**, $a(b + c) = ab + ac$. Multiply **each** term of the polynomial by the monomial. Remember the following when completing your work.

Helpful Hints
- The multiplication can be arranged either horizontally or vertically.
- Use the distributive property to multiply each term of the polynomial by the monomial.
- Be careful with the signs.
- Be careful with exponents.

Multiply each of the following.

1) **Multiply $(5a^2 - 7a + 9)$ by 2a using the horizontal method.**

$2a (5a^2 - 7a + 9)$ *Multiply each term by 2a.*

$= 10a^3 - 14a^2 + 18a$

2) **$-6ab (2a^2 - 4ab + 5b^2)$**   *Multiply each term by -6ab. Be careful of signs.*

$-12a^3b + 24a^2b^2 - 30ab^3$

3) **Multiply $(x^2 + 5x - 4)$ by 3x using the vertical method.**

$$\begin{array}{r} x^2 + 5x - 4 \\ \underline{\times \quad 3x} \\ 3x^3 + 15x^2 - 12x \end{array}$$

Multiply each term by 3x.

4)

$$\begin{array}{r} 3xz - 6xy - 2yz \\ \underline{\times \quad -5xyz} \\ -15x^2yz^2 + 30x^2y^2z + 10xy^2z^2 \end{array}$$

Be careful of signs.

6-5 Multiplying Polynomials by Monomials

For 1–10 multiply using the vertical method.

1) $(2bc + 5ab)$ and $4c$

2) $(-9ac + 2c - 4c)$ and $-5c$

3) $(2m^2 - 5n^3 + 3m^2)$ and $10m^4 n$

4) $(-3x^2 + xy - z^2)$ and $-5xyz$

5) $(2x^2 - 7x + 7)$ and $-5x$

6) $(8x + 7y)$ and -4

7) $3x$ and $(2 - 5x + 5x^2)$

8) $7m$ and $(3m^2 - 7m + 3)$

9) $-3x$ and $(4x^2 - 7x + 3)$

10) $3x^2 y^2$ and $(-4x^2 + 2xy - 2x)$

For 11–20 multiply using the horizontal method.

11) $(3x^3 - 2x^2 + 1)$ and $4x$

12) $(6x^3 - 2xy - 4y)$ and $-3xy$

13) $(y^2 - 3yz + z^2)$ and $4yz$

14) $(8ab^2 - 3b^2)$ and $-3a^2 b^2$

15) $(-2r^2 + 4rs - 3s^2)$ and $3r^2 s^2$

16) $6x^2 y$ and $(2x - 5y)$

17) x^3 and $(5 - 7x + 3x^2 - x^4)$

18) -20 and $(3x^2 - xy + 4)$

19) $-3x$ and $(x^2 y^2 - 4xy + x^3)$

20) $2xy$ and $(x^3 - 3xy + y^2)$

Multiply each expression.

1) $(xy)(x^2 y)$

2) $(3x^2)(-2x^4)$

3) $(5x^2)(4xy^2)$

4) $(-6a^2 b^5)(a b c^3)$

6-6 Multiplying Binomials

INTRODUCTION

The distributive property is used when multiplying two binomials. It is easiest to multiply them horizontally. A useful method is sometimes called the **FOIL** method. **FOIL** stands for **First, Outer, Inner,** and **Last**. With practice it isn't necessary to write out each step. Remember the following steps when using the FOIL method of multiplying two binomials.

Helpful Hints
- **First**, multiply the **first** terms.
- **Second**, multiply the **outer** terms and then multiply the **inner** terms. Then add the two answers. Be careful with signs. With practice, this step can be done quickly in your head.
- **Third**, multiply the **last** two terms.
- **Fourth**, write out the result. Just remember FOIL.

EXAMPLES

Multiply each pair of binomials.

1) $(x + 4) (x + 5)$

First $(x + 4) (x + 5) = x^2$ *Multiply the first terms*

Second $(x + 4) (x + 5) = 5x + 4x = 9x$ *Multiply the outer terms. Multiply the inner terms. Add the answers.*

Third $(x + 4) (x + 5) = 20$ *Multiply the last terms*

Add the results $x^2 + 9x + 20$ *is the answer*

Remember, as you get better, you can do the steps mentally.

2) $(2x + 3) (3x - 2)$

First $(2x + 3) (3x - 2) = 6x^2$ *First*

Second $(2x + 3) (3x - 2) = -4x + 9x = 5x$ *Outer, Inner, Add*

Third $(2x + 3) (3x - 2) = -6$ *Last*

Add the results $6x^2 + 5x - 6$ *is the answer*

6-6 Multiplying Binomials

Multiply the following pairs of binomials. As you feel more comfortable, begin to do the steps mentally without showing all the work.

1) $(x + 5)(x + 3)$

2) $(x + 3)(x - 4)$

3) $(x - 3)(x - 5)$

4) $(3x + 2)(x + 5)$

5) $(x - 6)(2x - 1)$

6) $(7x + 3)(2x - 1)$

7) $(5x - 3)(2x - 1)$

8) $(2x + 3)(3x + 2)$

9) $(-5x - 4)(-5x - 4)$

10) $(2x + 3)(x - 2)$

11) $(-2x + 1)(2x + 2)$

12) $(3x + 2)(x + 4)$

13) $(2x + 3y)(3x + 4y)$

14) $(5x - 3y)(2x - y)$

15) $(6m - 3n)(2m - n)$

16) $(5x - 2y)(2x + 3y)$

REVIEW

Divide each of the following.

1) $\dfrac{8x^3}{2x}$

2) $\dfrac{-32x^4}{8x^2}$

3) $\dfrac{-24a^7 b^6}{8a^3 b^2}$

4) $\dfrac{-6xy}{xy}$

6-7 Dividing Polynomials by Monomials

To divide a polynomial by a monomial, divide each term of the polynomial by the monomial. Remember the following.

Helpful Hints

- **First**, arrange the terms of the polynomial in descending order if possible.
- **Second**, take each term one at a time and divide.
- Use the laws of exponents when possible. It might be good to review Lesson 1-8 Laws of Exponents.

EXAMPLES

Divide each of the following.

1) $$\frac{6xy^2 + 9x^2y}{3xy}$$

$$= \frac{6xy^2}{3xy} + \frac{9x^2y}{3xy}$$ *Divide each term by 3xy and simplify.*

$$= 2y + 3x$$

2) $$\frac{8m^2n - 4mn^2 + 16mn}{4mn}$$

$$= \frac{8m^2n}{4mn} - \frac{4mn^2}{4mn} + \frac{16mn}{4mn}$$ *Divide each term by 4mn and simplify.*

$$= 2m - n + 4$$

3) $$\frac{6x^5 - 21x^4 + 18x^3 - 3x^2 + 9x}{3x}$$

$$= \frac{6x^5}{3x} - \frac{21x^4}{3x} + \frac{18x^3}{3x} - \frac{3x^2}{3x} + \frac{9x}{3x}$$ *Divide each term by 3x and simplify.*

$$= 2x^4 - 7x^3 + 6x^2 - x + 3$$

6-7 Dividing Polynomials by Monomials

Divide each of the following.

1) $\dfrac{6x + 2x}{2x}$

2) $\dfrac{4x^2 + 6x - 6xy}{2x}$

3) $\dfrac{y^3 + y^2z + yz^2}{y}$

4) $\dfrac{6x^3 + 9x^2 - 3x}{3x}$

5) $\dfrac{6x^2y + 9xy^2 - 18x}{3x}$

6) $\dfrac{6xy^2 + 4xy - 6x^2y}{2xy}$

7) $\dfrac{14x^3y^4 - 21x^5y^3 + 35x^2y^5}{7x^2y^3}$

8) $\dfrac{21x^4y^3 - 9x^5y^4 + 21x^3y^2}{3xy}$

9) $\dfrac{-18x^6y^3 + 12x^4y^2 - 9x^2y^3}{-3x^2y^2}$

10) $\dfrac{8x^4y^4 + 4x^3y^3 + 4x^2y^2 + 4xy}{2xy}$

11) $\dfrac{14x + 21y - 35}{-7}$

12) $\dfrac{14x^2y - 21xy^2 + 7x^3y}{7xy}$

13) $\dfrac{x^3 - x^2 + 2x}{-x}$

14) $\dfrac{9x^2y^3 - 6x^2y^2 + 15x^3y^2}{3x^2y^2}$

15) $\dfrac{33x^3 - 66x^4 + 22x^2}{11x^2}$

16) $\dfrac{15x^3y^3 - 25x^2y^2 + 30x^3y^2}{5x^2y^2}$

REVIEW

Multiply or divide each of the following.

1) $(3x^3)(-4x^4)$

2) $\dfrac{x^3y^2}{x^2y}$

2) $(3a^2b^2)(a^3b^3c)$

4) $\dfrac{14x^4y^3}{-2xy^2}$

6-8 Dividing a Polynomial by a Binomial

INTRODUCTION

Dividing a polynomial by a binomial is similar to the long division you learned in elementary school arithmetic. Just as in long division some answers will have no remainder while other answers will have a remainder. Remember the following.

Helpful Hints
- **First**, set the problem up just like you would an arithmetic long division problem.
- Remember the basic steps of long division:
  Divide, Multiply, Subtract, Divide again.
- The answer is called a **quotient**. It is best to express your remainder as a fraction.

EXAMPLES

Find the quotient of
$(x^2 - x - 8) \div (x + 2)$

$$
\begin{array}{r}
x \\
x + 2 \overline{\smash{)}x^2 - x - 8}
\end{array}
$$
Divide

$$
\begin{array}{r}
x \\
x + 2 \overline{\smash{)}x^2 - x - 8} \\
x^2 + 2x
\end{array}
$$
Multiply

$$
\begin{array}{r}
x \\
x + 2 \overline{\smash{)}x^2 - x - 8} \\
\underline{x^2 + 2x} \\
-3x
\end{array}
$$
Subtract -x – 2x = -3x

$$
\begin{array}{r}
x - 3 \\
x + 2 \overline{\smash{)}x^2 - x - 8} \\
\underline{x^2 + 2x} \\
-3x - 8
\end{array}
$$
Bring down the 8 and divide.

$$
\begin{array}{r}
x - 3 \\
x + 2 \overline{\smash{)}x^2 - x - 8} \\
\underline{x^2 + 2x} \\
-3x - 8
\end{array}
$$
Multiply

$$
\begin{array}{r}
x - 3 \\
x + 2 \overline{\smash{)}x^2 - x - 8} \\
\underline{x^2 + 2x} \\
-3x - 8 \\
\underline{-3x - 6} \\
-2
\end{array}
$$
Subtract -8 – (-6) = -8 + 6 = -2

The answer is $x - 3 + \dfrac{-2}{x + 2}$

Quotient   *Remainder*

6-8 Dividing a Polynomial by a Binomial

Find the quotient of each of the following. If there is a remainder, include it in the answer.

1) $(x^2 + 4x + 4) \div (x + 2)$ 2) $(x^2 + 10x + 24) \div (x + 6)$

3) $(x^2 - 4x - 14) \div (x - 6)$ 4) $(x^2 - x - 12) \div (x + 3)$

5) $(x^2 - 9x + 20) \div (x - 3)$ 6) $(x^2 - 9x + 7) \div (x - 2)$

7) $(x^2 - 8x + 9) \div (x - 1)$ 8) $(x^2 - 15x - 59) \div (x + 3)$

9) $(3x^2 - 8x + 4) \div (3x - 2)$ 10) $(21x^2 - 10x + 1) \div (3x - 1)$

11) $(6x^2 - 11x - 8) \div (3x + 2)$ 12) $(x^2 - 3x + 2) \div (x - 2)$

13) $(x^2 + 3x + 2) \div (x + 2)$ 14) $(3n^2 - n - 2) \div (3n + 2)$

15) $(3x^2 + 14x + 10) \div (x + 4)$ 16) $(15x^2 + 17x - 4) \div (3x + 4)$

REVIEW

Add or subtract each of the following.

1) $(4x^4 + 7x^3 + 15x^2 + 4) + (4x^3 + 2x^2 + 17x)$

2) $(m^4 + m^3 - 2m^2 + 7m - 5) - (-5m^4 - 2m^3 - 4m^2 - m)$

6-9 Factoring Using Common Monomial Factors

INTRODUCTION

When we factor a polynomial it is good to look first for the **greatest common monomial factor**. We are looking for factors that are common to all terms in the polynomial. With practice, it becomes easy to recognize these common monomial factors. Remember the following steps when factoring a polynomial whose terms have a common monomial factor.

Helpful Hints
- **First**, find the greatest monomial that is a factor of each of the terms of the polynomial.

- **Second**, divide each term of the polynomial by the monomial factor. The answer is the other factor.

- **Third**, rewrite the polynomial as the product of the two factors.

EXAMPLES

Factor each of the following.

1) $3x^2 + 6x + 15$

$= 3(x^2 + 2x + 5)$

The greatest common monomial factor is 3.

After dividing each term by 3, we get our answer.

2) $16a^2b + 20ab^2 + 2ab$

$= 2ab(8a + 10b + 1)$

The greatest common monomial factor is 2ab.

After dividing each term by 2ab, we get our answer.

3) $15x^3 - 20x^2$

$= 5x^2(3x - 4)$

The greatest common monomial factor is $5x^2$.

After dividing each term by $5x^2$, we get our answer.

4) $8r^4s^2t^3 + 12r^2s^2t^2$

$= 4r^2s^2t^2(4r^3t + 3)$

The greatest common monomial factor is $4r^2s^2t^2$.

After dividing each term by $4r^2s^2t^2$, we get our answer.

6-9 Factoring Using Common Monomial Factors

Factor each of the following.

1) $28x^2 - 7x$

2) $4x^2 - 8x$

3) $12x^2 + 48x^2y$

4) $21n^2 - 14mn$

5) $6x^2 + 12x + 24xy + 36$

6) $32a^2b^4 - 16ab^3 + 40a^3b^5$

7) $21mn + 28m^2n^2 + 35m^3n^3$

8) $4a^2b^7 - 32a^2b^6 + 8ab^2$

9) $21x^3y^2 - 7x^2y + 42xy$

10) $5a^4 + 25a^3 - 35a^2 + 20a$

11) $3x^4y - 18y^3$

12) $4x^5y^3 + 4x^3y^3$

13) $3xy^3 - 15xy^2 - 18xy$

14) $x^3 + x^2 + 2x$

15) $2x^2 + 8x + 4$

16) $x^3 - 4x^2 - 4x$

17) $15x^3 - 9x^2 + 3x$

18) $16x^3 + 25xy^2$

19) $5x^2 + 50x^3 - 125x$

20) $3x^3 + 30x^2 - 75x$

REVIEW

Multiply.

1) $4m(m + 2n + 3)$

2) $-5x(3x + 2y)$

3) $-x^2(x + 2x^2)$

4) $-2c(2c^2 + 4c - 5)$

6-10 Factoring Trinomials of the Form $x^2 + bx + c$

INTRODUCTION

When factoring a trinomial in the form of **$x^2 + bx + c$** the result will be **two binomials**. This is process is the reverse of multiplying two binomials. Remember the following. With practice, these become easy.

Helpful Hints
- The product of the first terms of both binomials is equal to the first term of the trinomial.
- The product of the second terms of both binomials is equal to the last term of the trinomial.
- To get the middle term of the trinomial, multiply the first term of each binomial by the second term of the other binomial. Then add the products.

EXAMPLES

Factor each of the following.

1) $x^2 + 5x + 6$

    The first two terms are both x.   (x   )   (x   )

    We want two factors whose **product** is **6** and whose **sum** is **5**. They would be **2 and 3**, since 2 x 3 = 6 and 2 + 3 = 5.

    Our answer is (x + 2) (x + 3).   *Multiply to check your answer.*

2) $x^2 - 8x + 15$

    The first two terms are both x.   (x   )   (x   )

    We want two factors whose **product** is **15** and whose **sum** is **-8**. They would be **-3 and -5**, since -3 x -5 = 15 and -3 + -5 = -8

    Our answer is (x – 3) (x – 5).   *Multiply to check your answer.*

3) $x^2 + 4x - 12$

    The first two terms are both x.   (x   )   (x   )

    We want two factors whose **product** is **-12** and whose **sum** is **4**. They would be **6 and -2**, since 6 x -2 = -12 and 6 + -2 = 4

    Our answer is (x + 6) (x – 2).   *Multiply to check your answer.*

4) $x^2 - 2x - 15$

The first two terms are both x.　(x　)　(x　)

We want two factors whose **product** is **-15** and whose **sum** is **-2**. They would be **-5 and 3**, since -5 x 3 = -15 and -5 + 3 = -2.

Our answer is (x – 5) (x + 3).　　*Multiply to check your answer.*

EXERCISES

Factor each of the following.

1) $x^2 + 4x + 3$　　　　2) $x^2 + 11x + 24$　　　　3) $x^2 + 12x + 32$

4) $x^2 + 24x + 63$　　　　5) $x^2 - 5x + 6$　　　　6) $x^2 - 12x + 27$

7) $x^2 - 2x - 8$　　　　8) $x^2 + 5x - 24$　　　　9) $x^2 - 2x - 48$

10) $x^2 + 15x + 26$　　　11) $x^2 - 4x - 32$　　　12) $x^2 - 13x + 40$

13) $x^2 + 4x - 5$　　　　14) $x^2 + 2x + 1$　　　　15) $x^2 - 2x - 8$

16) $x^2 - 16x + 48$　　　17) $x^2 - 12x - 13$　　　18) $x^2 + 3x - 40$

REVIEW

Multiply using the foil method.

1) (x + 2) (x + 3)　　　　　　2) (2x + y) (x + 2y)

3) (y – 5) (y + 4)　　　　　　4) (3y – 2) (2y + 1)

6-11 Factoring the Difference of Two Squares

An expression in the form of $a^2 - b^2$ is called the **difference of two squares**. It is often necessary to factor the difference of two squares. With practice, it is easy to get good at recognizing them.

It is important to keep in mind that $a^2 - b^2 = (a + b)(a - b)$. Remember the following when factoring the difference of two squares.

Helpful Hints
- The answer will be two binomials.
- **First**, get the values of the first terms and second terms of the binomials.
- **Second**, connect the first pair with a plus (+) sign and the second pair with a minus (-) sign.

EXAMPLES

Factor each of the following.

1) $x^2 - 4y^2$ The first term would be **x**, since $x \cdot x = x^2$.

   The second term would be **2y**, since $2y \cdot 2y = 4y^2$.

   $(x + 2y)(x - 2y)$ is the answer.

2) $144 - 121x^2$ The first term would be **12**, since $12 \cdot 12 = 144$.

   The second term would be **11x**, since $11x \cdot 11x = 121x^2$.

   $(12 + 11x)(12 - 11x)$ is the answer.

3) $64x^2 - 9y^2$ The first term would be **8x,** since $8x \cdot 8x = 64x^2$.

   The second term would be **3y**, since $3y \cdot 3y = 9y^2$.

   $(8x + 3y)(8x - 3y)$ is the answer.

4) $81a^8b^4 - c^8$ The first term would be **$9a^4b^2$,** since $9a^4b^2 \cdot 9a^4b^2 = 81a^8b^4$.

   The second term would be **c^4**, since $c^4 \cdot c^4 = c^8$.

   $(9a^4b^2 + c^4)(9a^4b^2 - c^4)$ is the answer.

6-11 Factoring the Difference of Two Squares

5) $16a^4 - 1$ The first term would be **4a²**, since $4a^2 \cdot 4a^2 = 16a^4$.

The second term would be **1**, since $1 \cdot 1 = 1$.

$= (4a^2 + 1)\,(4a^2 - 1)$ Factor the difference of two squares.

$= (4a^2 + 1)\,(2a + 1)\,(2a - 1)$ This is the final answer.

EXERCISES

Factor each of the following.

1) $x^2 - 121$

2) $4x^2 - 25y^2$

3) $4y^2 - 1$

4) $100y^2 - a^2$

5) $16y^2 - 81$

6) $x^2 - y^2$

7) $144 - x^2$

8) $64x^2 - 4y^2$

9) $100x^2 - 81y^2$

10) $1 - x^2$

11) $16x^2 - 25y^2$

12) $121 - m^2$

13) $25x^4 - y^4$

14) $m^2 n^2 - 121$

15) $25m^6 - 16y^6$

16) $36a^2 b^2 - x^2 y^2$

17) $16m^2 n^2 - 9x^2 y^2$

18) $36x^6 - 25y^8$

REVIEW

Divide each of the following.

1) $\dfrac{5a + 15b}{5}$

2) $\dfrac{8x - 44}{2}$

3) $\dfrac{3x^4 + 3x^3 - 6x^2}{3x}$

4) $\dfrac{5a^2 + 45a}{5a}$

6-12 Factoring Using Combinations

Sometimes an expression can be factored several times. It is often necessary to look very closely at the terms to decide the easiest ways to factor the expression completely. Use the following steps when factoring an expression.

Helpful Hints
- **First**, check for a common monomial factor. If there is one, factor the expression.
- **Second**, check to see whether one of the factors is the difference of two squares. If so, factor using the rule $a^2 - b^2 = (a + b)(a - b)$
- **Third**, check to see if one of the factors is a trinomial, and factor it if possible.
- **Fourth**, write your answer as the product of all the factors.

EXAMPLES

Factor each of the following completely.

1) $2x^2 - 32$ *Each term has a factor of 2*
   $= 2(x^2 - 16)$ *Now factor ($x^2 - 16$), the difference of two squares*
   $= 2(x + 4)(x - 4)$

2) $2x^2 - 16x + 30$ *Each term has a factor of 2*
   $= 2(x^2 - 8x + 15)$ *Now factor $x^2 - 8x + 15$*
   $= 2(x - 5)(x - 3)$

3) $x^4 - 16$ *Factor the difference of two squares*
   $= (x^2 + 4)(x^2 - 4)$ *Factor the difference of two squares*
   $= (x^2 + 4)(x + 2)(x - 2)$

4) $x^3 - 4x^2 + 4x$ *Each term has a factor of x*
   $= x(x^2 - 4x + 4)$ *Now factor $x^2 - 4x + 4$*
   $= x(x - 2)(x - 2)$ *Simplify*
   $= x(x - 2)^2$

6-12 Factoring Using Combinations

Factor each of the following completely.

1) $xy^2 - xz^2$

2) $2x^2 - 4x - 48$

3) $x^3 + 7x^2 + 10x$

4) $2x^2 - 200y^2$

5) $18x^2 - 8$

6) $3x^2 + 30x + 27$

7) $2ax^2 - 2ax - 12a$

8) $4a^2 - 36$

9) $3x^2 - 48$

10) $4x^2 - 40x + 100$

11) $4x^2 - 6x - 4$

12) $y^3 - 25y$

13) $x^3 + 7x^2 + 10x$

14) $2x^2 - 32$

15) $4x^2 - 4x - 48$

16) $mx^2 - my^2$

17) $2x^2 + 6x - 8$

18) $2a^2y^2 - 3a^2y + a^2$

REVIEW

Divide each of the following.

1) $(x^2 + 6x + 8) \div (x + 4)$

2) $\dfrac{6x^2 + 15xy^2 + 6x^2y^2}{3x}$

6-13 Factoring More Difficult Trinomials

When the first term of a trinomial has a coefficient that is not 1, factoring can be more difficult. The trinomial $3x^2 + 7x + 2$ is an example of this. These have to be worked using the trial and error method. Use the following steps when factoring more difficult trinomials.

Helpful Hints
- **First**, factor the first term.
- **Second**, find all the possible factors of the last term.
- **Third**, choose the factors of the last term that will result in the middle term of the trinomial.
- Remember, sometimes a lot of trial and error is involved. With practice they get much easier.

EXAMPLES

Factor each of the following.

1) $2x^2 - 3x - 9$

The first two terms are 2x and x.

$(2x \quad)(x \quad)$

The factors of -9 are   **1** and **-9**,   **-1** and **9**,   **3** and **-3**

Choose the pair that will result in the middle term -3x.

Our answer is $(2x + 3)(x - 3)$ *Multiply to check your answer.'*

2) $3x^2 + 7x + 2$

The first two terms are 3x and x.

$(3x \quad)(x \quad)$

The factors of 2 are   **2** and **1**,   **-2** and **-1**

Choose the pair that will result in the middle term 7x.

Our answer is $(3x + 1)(x + 2)$ *Multiply to check your answer.*

Some trinomials can be quite difficult to factor.
Now you have a strategy to work with.

6-13 Factoring More Difficult Trinomials

Factor each of the following.

1) $2x^2 + 5x + 2$

2) $3x^2 + 10x + 8$

3) $2x^2 - 3x + 1$

4) $3x^2 - 8x + 4$

5) $3x^2 - 5x - 12$

6) $5x^2 - 3x - 8$

7) $2x^2 + x - 6$

8) $2x^2 + 5x - 3$

9) $3x^2 + x - 2$

10) $2x^2 - 3x - 9$

11) $2x^2 + 11x + 5$

12) $2x^2 + 11x + 12$

13) $5x^2 - 16x + 3$

14) $3x^2 - x - 14$

15) $3x^2 - 11x + 10$

16) $7x^2 - 9x + 2$

17) $2x^2 + 11x + 15$

18) $4x^2 + 2x - 12$

REVIEW

Multiply each of the following.

1) $(x + 3)(x + 2)$

2) $(x + 7)(x - 4)$

3) $(2x + 1)(x - 4)$

4) $(2x - 3)(3x + 2)$

Chapter 6 Review: Polynomials

Simplify each of the following.

1) $(12x^2 + 7xy + 5x) + (13x^2 - 5xy + 7x)$

2) $(15x^3 + 7x^2 + 5x + 3) - (8x^3 - 5x^2 + 3x - 5)$

3) $5x^2y \cdot 7x^3y^2$

4) $(2xy)(3x^2y^2)(2x^2y^3)$

5) $\frac{15a^5}{3ab^2}$

6) $xy(x^3 + 3xy - 1)$

7) $(3x - 7)(2x + 3)$

8) $\frac{35x^2 + 5x}{5x}$

Factor each of the following.

9) $24x - 8x^2$

10) $x^2 + 10x + 25$

11) $2x^2 + 12x + 16$

12) $x^2 - 49$

13) $49x^2 - 81y^2$

14) $98x^2 - 200y^2$

15) $x^3y - xy^3$

16) $a^2b^2 + 12a^2b + 36a^2$

17) $3x^2 - 10x - 8$

18) $6x^2 - 7x - 3$

Are You Getting the Most Possible From This Book?

Here are a list of questions that you should ask yourself to determine whether you are taking the actions necessary to get the most possible out of using this book.

- Are you viewing each Online Tutorial Video lesson, and working along with the instructor?

- Are you carefully reading the Introduction to each lesson?

- Are you carefully reading the Helpful Hints section for each lesson?

- Are you neatly copying each Example and carefully writing down all the steps? Doing this will help you to more effectively understand the problem, and you will be more prepared for the Exercises.

- Are you neatly and carefully completing all of the Exercises, showing the work for each problem?

- Are you using the Solutions section to correct your work?

- Are you re-working the problems that you worked incorrectly, to find out what caused the mistake?

- Are you occasionally reviewing chapters that you have completed?

- Are you using the Glossary and other resources located in the back of the book? Reading through the Glossary is a good way to review terms and their definitions.

Every question that you answered with yes represents an important step towards effectively learning algebra. It is up to you to use this book in a way that will benefit you the most!

7-1 Simplifying Algebraic Fractions

Algebraic fractions are very similar to fractions that you learned in elementary school. The difference is that algebraic fractions contain variables as well as numerals. You can simplify an algebraic fraction by dividing the numerator and the denominator by the same number. This is really just cancelling out common factors.

It might be a good idea to review the laws of exponents.

When a fraction has been simplified, the value is not changed. The fraction is just written in a simpler way. A fraction is in its simplest form when the numerator and the denominator have no common factors except for 1. It is important to remember that the **denominator** in a fraction **cannot be equal to zero (0)**.

Remember the following when simplifying algebraic fractions.

Helpful Hints
- Cancel out any common factors found in the numerator and denominator.
- Be careful of signs.
- Express your answer with positive exponents.
- Remember $\dfrac{a^m}{a^n} = a^{m-n}$
- Remember $a^{-n} = \dfrac{1}{a^n}$
- Keep in mind that algebraic fractions are quite similar to arithmetic fractions.

EXAMPLES

Simplify each of the following. Use positive exponents in your answers.

1) $\dfrac{12a^2}{4a}$

$= \dfrac{12}{4} \cdot a^{2-1}$

$= 3a^1$

$= 3a$

2) $\dfrac{6y^2 z}{-3y}$

$= \dfrac{6}{-3} \cdot y^{2-1} \cdot z$

$= -2 \cdot y^1 \cdot z$

$= -2yz$

3) $\dfrac{10x^2 y^8}{35x^7 y^2}$

$= \dfrac{10}{35} \cdot x^{2-7} \cdot y^{8-2}$

$= \dfrac{2}{7} \cdot x^{-5} \cdot y^6$

$= \dfrac{2y^6}{7x^5}$

4) $\dfrac{125a^2 b^4 c}{-25a^3 b\, c^3}$

$= \dfrac{125}{-25} \cdot a^{2-3} \cdot b^{4-1} \cdot c^{1-3}$

$= -5a^{-1} \cdot b^3 \cdot c^{-2}$

$= \dfrac{-5b^3}{a^1 c^2} = \dfrac{-5b^3}{ac^2}$

7-1 Simplifying Algebraic Fractions

Simplify each of the following. Express you answer with positive exponents.

1) $\dfrac{10m^2}{5m}$

2) $\dfrac{5a^2 b^3}{15a^3 b^2}$

3) $\dfrac{-12x^2 y^2}{3x^3 y^3}$

4) $\dfrac{12n^2 r}{4n\, r^2}$

5) $\dfrac{10x^2 y^2 z^2}{-6x\, y^4 z^8}$

6) $\dfrac{9mn^2}{8n^2}$

7) $\dfrac{20a^2 b^4}{5a^3 b}$

8) $\dfrac{-9x^1 y^2 z^3}{-3x^2 y^3 z^4}$

9) $\dfrac{72xy}{8x^2 y^3}$

10) $\dfrac{r\, s\, t}{r^3 t^2}$

11) $\dfrac{15x^2 y}{xy}$

12) $\dfrac{3xy}{15x^3 y^4}$

13) $\dfrac{32x^2 y^2 z^2}{12xy^2 z^2}$

14) $\dfrac{-12a^2 b^3 c^2}{2a^3 b^4 c}$

15) $\dfrac{-25x^2 y^3}{35x^3 y^2}$

16) $\dfrac{7xy}{28x^3 y^3}$

17) $\dfrac{27m^2 n^2 r^3}{9m^4 n^3 r^2}$

18) $\dfrac{24x\, y^2 z^2}{3x^4 y^5 z^5}$

REVIEW

Factor completely each of the following.

1) $2x^2 + 14x + 20$

2) $4x^2 - 40x + 100$

3) $8x^2 - 32$

4) $2x^2 - 162$

7-2 Simplifying Algebraic Fractions with Several Terms

INTRODUCTION

Sometime algebraic fractions have several terms in the numerator or denominator. They can be simplified also. Just remember that the approach is basically the same as when you simplified arithmetic fractions. Keep the following in mind when simplifying algebraic fractions with several terms.

Helpful Hints
- **First**, if possible, factor the numerator.
- **Second**, if possible, factor the denominator.
- **Third**, cancel out any common factors.
- Remember that an algebraic fraction is in its simplest form when the numerator and denominator have no common factors except 1.
- Review the laws of exponents if necessary.

EXAMPLES

Simplify each of the following. Factor the numerator and denominator when possible.

1) $\dfrac{x^2 + 2x}{x + 2}$

$= \dfrac{x\,(x + 2)}{(x + 2)}$

$= x$

2) $\dfrac{x^2 - 2x + 1}{x^2 - x}$

$= \dfrac{(x - 1)\,(x - 1)}{x\,(x - 1)}$

$= \dfrac{x - 1}{x}$

3) $\dfrac{x^2 + 6x + 9}{x^2 - 9}$

$= \dfrac{(x + 3)\,(x + 3)}{(x + 3)\,(x - 3)}$

$= \dfrac{x + 3}{x - 3}$

4) $\dfrac{x^2 - 12x + 20}{x^2 + 4x - 12}$

$= \dfrac{(x - 2)\,(x - 10)}{(x - 2)\,(x + 6)}$

$= \dfrac{x - 10}{x + 6}$

5) $\dfrac{x^2 - y^2}{3x - 3y}$

$= \dfrac{(x + y)\,(x - y)}{3(x - y)}$

$= \dfrac{x + y}{3}$

6) $\dfrac{5x^2 - 20}{(x - 2)^2}$

$= \dfrac{5(x^2 - 4)}{(x - 2)^2}$

$= \dfrac{5(x + 2)\,(x - 2)}{(x - 2)\,(x - 2)}$

$= \dfrac{5(x + 2)}{x - 2}$

7-2 Simplifying Algebraic Fractions with Several Terms

EXERCISES

Simplify each of the following.

1) $\dfrac{x + 1}{x^2 + x}$

2) $\dfrac{3x}{9x^2 - 6x}$

3) $\dfrac{x^2 - y^2}{x + y}$

4) $\dfrac{2m - 4mn}{1 - 2n}$

5) $\dfrac{(x - 4)^2}{3x - 12}$

6) $\dfrac{6x + 6y}{x^2 - y^2}$

7) $\dfrac{2x^2 - 50}{2x + 10}$

8) $\dfrac{x^2 - x - 6}{x^2 - 9}$

9) $\dfrac{x^2 - 5x + 6}{x^2 - x - 6}$

10) $\dfrac{x^2 - 4x - 5}{x^2 - 2x - 15}$

11) $\dfrac{x^2 - x - 6}{x^2 + 5x + 6}$

12) $\dfrac{x^2 + 7x + 10}{x^2 - x - 6}$

13) $\dfrac{(x + y)^2}{x^2 - y^2}$

14) $\dfrac{9y - 18}{3y^2 - 12}$

15) $\dfrac{x^2 + 5x + 6}{x^2 - 9}$

16) $\dfrac{2x^2 - 7x + 3}{(x - 3)^2}$

17) $\dfrac{(x + y)^2}{x^2 - y^2}$

18) $\dfrac{x^2 - 6x}{x^2 - 7x + 6}$

REVIEW

Factor each difference of two squares.

1) $x^2 - 81$

2) $m^2 - 16m^2$

3) $144 - 121x^2$

4) $36b^2 - 4a^2$

7-3 Using the -1 Factor to Simplify Algebraic Fractions

Using **-1** as a factor can sometimes be a very useful tool when simplifying algebraic fractions. Keep in mind that $(a - b) = -1(b - a)$. This is something that you need to remember because it can prevent a lot of unnecessary difficulties. Remember the following when using the -1 factor to simplify algebraic fractions.

Helpful Hints

- Factor the numerators and denominators as usual.
- If you have a fraction that is similar to $\frac{x - 2}{2 - x}$ you can use the -1 factor to simplify.
- Keep in mind that there are three ways to write a negative fraction.

  Example: $-\frac{1}{x} = \frac{-1}{x} = \frac{1}{-x}$. They all mean the same thing.

EXAMPLES

Simplify each of the following using the -1 factor.

1) $\dfrac{x - 1}{1 - x}$

$= \dfrac{x - 1}{-1(x - 1)}$ *Use the -1 factor*

$= \dfrac{1}{-1}$

$= -1$

2) $\dfrac{x - y}{3y - 3x}$

$= \dfrac{x - y}{3(y - x)}$ *Factor*

$= \dfrac{-1(y - x)}{3(y - x)}$ *Use the -1 factor*

$= \dfrac{-1}{3}$

$= -\dfrac{1}{3}$

3) $\dfrac{x^2 - y^2}{3y - 3x}$

$= \dfrac{(x + y)(x - y)}{3(y - x)}$ *Factor*

$= \dfrac{(x + y)(x - y)}{3 \cdot -1(x - y)}$ *Use the -1 factor*

$= \dfrac{x + y}{-3}$

$= -\dfrac{x + y}{3}$

4) $\dfrac{8 - 4x}{x^2 + x - 6}$

$= \dfrac{4(2 - x)}{(x + 3)(x - 2)}$ *Factor*

$= \dfrac{4 \cdot -1(x - 2)}{(x + 3)(x - 2)}$ *Use the -1 factor*

$= \dfrac{4 \cdot -1}{x + 3}$

$= \dfrac{-4}{x + 3}$

$= -\dfrac{4}{x + 3}$

7-3 Using the -1 Factor to Simplify Algebraic Fractions

Simplify each of the following using the -1 factor.

1) $\dfrac{x-4}{4-x}$

2) $\dfrac{m-n}{n-m}$

3) $\dfrac{1-x}{2x-2}$

4) $\dfrac{m^2-9}{3-m}$

5) $\dfrac{3x^2-3}{3-3x}$

6) $\dfrac{1-x}{x^2-1}$

7) $\dfrac{5-5m}{m^2+3m-4}$

8) $\dfrac{2x-2y}{y^2-x^2}$

9) $\dfrac{m-3n}{3n-m}$

10) $\dfrac{4x-4y}{2y-2x}$

11) $\dfrac{5m-5n}{5n-5m}$

12) $\dfrac{n-7}{7-n}$

13) $\dfrac{x^2-y^2}{3y-3x}$

14) $\dfrac{3-x}{x^2-9}$

15) $\dfrac{2-x}{4x-8}$

16) $\dfrac{3x-3}{1-x^2}$

17) $\dfrac{3x-9}{9-x^2}$

18) $\dfrac{x^2-y^2}{y-x}$

REVIEW

Factor each of the following.

1) $x^2+9x+14$

2) x^2-4x+4

3) $x^2-3x-40$

4) $x^2-5x-36$

7-4 Solving Proportions Containing Algebraic Fractions

INTRODUCTION

When two algebraic fractions are equal it is called a **proportion**. Sometimes the proportion may look complicated, but to solve it, simply cross-multiply. The next step is to then solve the equation. Keep the following in mind when **solving proportions**.

Helpful Hints
- **First**, cross multiply.
- **Second**, try to isolate the variable on the left side of the equal sign.
- It is often necessary to use the distributive property when cross-multiplying.

EXAMPLES

Solve each of the following proportions.

1) $\dfrac{3x}{5} = \dfrac{6}{5}$

$\dfrac{3x}{5} \diagdown \dfrac{6}{5}$ *Cross-multipy*

$5 \cdot 3x = 5 \cdot 6$

$15x = 30$

$\dfrac{15x}{15} = \dfrac{30}{15}$ *Divide both sides by 15*

$x = 2$

2) $\dfrac{3x-5}{8x} = \dfrac{1}{4}$

$\dfrac{3x-5}{8x} \diagdown \dfrac{1}{4}$ *Cross multiply*

$4(3x - 5) = 8x$ *Use the distributive property*

$12x - 20 = 8x$

$12x - 8x - 20 = 8x - 8x$ *Subtract 8x from both sides*

$4x - 20 = 0$

$4x - 20 + 20 = 20$ *Add 20 to both sides*

$4x = 20$

$x = 5$

3) $\dfrac{x+3}{5} = \dfrac{x+1}{4}$

$\dfrac{x+3}{5} \diagdown \dfrac{x+1}{4}$ *Cross multiply*

$5(x + 1) = 4(x + 3)$ *Use the distributive property*

$5x + 5 = 4x + 12$

$5x - 4x + 5 = 4x - 4x + 12$ *Subtract 4x from both sides*

$x + 5 = 12$

$x = 7$

4) $\dfrac{x}{x-2} = \dfrac{x+2}{x+4}$

$\dfrac{x}{x-2} \diagdown \dfrac{x+2}{x+4}$ *Cross multiply*

$x(x + 4) = (x + 2)(x - 2)$ *Multiply*

$x^2 + 4x = x^2 - 4$

$x^2 - x^2 + 4x = x^2 - x^2 - 4$ *Subtract x² from both sides*

$4x = -4$

$x = -1$

Chapter 7: **RATIONAL EXPRESSIONS**

7-4 Solving Proportions Containing Algebraic Fractions

Solve each of the following proportions.

1) $\frac{3n}{4} = \frac{3}{2}$

2) $\frac{9}{x} = \frac{6}{8}$

3) $\frac{x-3}{8} = \frac{3}{4}$

4) $\frac{4}{2x-1} = \frac{3}{x}$

5) $\frac{y+2}{y-2} = \frac{10}{6}$

6) $\frac{2n+3}{4n} = \frac{6}{8}$

7) $\frac{n+6}{6} = \frac{n+5}{4}$

8) $\frac{5}{x+2} = \frac{4}{x}$

9) $\frac{3x+3}{3} = \frac{7x-1}{5}$

10) $\frac{x}{x+4} = \frac{x+1}{x+6}$

11) $\frac{x-3}{x+5} = \frac{x-4}{x+3}$

12) $\frac{x-3}{x-5} = \frac{x+5}{x-1}$

13) $\frac{x+10}{x} = \frac{9}{6}$

14) $\frac{7x-1}{5} = \frac{3x+3}{3}$

15) $\frac{x-2}{4} = \frac{x+10}{10}$

16) $\frac{x+4}{3} = \frac{x}{5}$

17) $\frac{14}{3} = \frac{7x}{2}$

18) $\frac{x-5}{2} = \frac{x+6}{3}$

REVIEW

Factor completely each of the following.

1) $x^2 + x - 6$

2) $9y^2 - 49$

3) $4x^2 + 8x - 60$

4) $4x^2 - 6x - 4$

7-5 Multiplying Algebraic Fractions

INTRODUCTION

Multiplying algebraic fractions is very similar to multiplying arithmetic fractions. Remember the following when multiplying algebraic fractions.

Helpful Hints

- **First**, factor the numerators and denominators of each fraction if possible.
- **Second**, carefully cancel out any common factors.
- **Third**, to get the final answer, multiply the remaining numerators and the remaining denominators.
- To make an integer or polynomial a fraction, simply place a 1 as its denominator.

Examples: $7 = \dfrac{7}{1}$ $4x + 2 = \dfrac{4x + 2}{1}$

EXAMPLES

Simplify each of the following.

1) $\dfrac{5x}{3y^2} \cdot \dfrac{6y^2}{x^2}$

$= \dfrac{5\cancel{x}}{_1 3\cancel{y^2}_1} \cdot \dfrac{^2\cancel{6}\cancel{y^2}^1}{\cancel{x^2}_x}$ *Cancel*

$= \dfrac{5 \cdot 2}{x}$ *Multiply*

$= \dfrac{10}{x}$

2) $\dfrac{x^2 - x}{5} \cdot \dfrac{25}{x}$

$= \dfrac{x(x - 1)}{5} \cdot \dfrac{25}{x}$ *Factor*

$= \dfrac{^1\cancel{x}(x - 1)}{_1\cancel{5}} \cdot \dfrac{^5\cancel{25}}{\cancel{x}_1}$ *Cancel*

$= 5(x - 1)$

3) $\dfrac{x^2 - y^2}{x^2 - 49} \cdot \dfrac{x + 7}{x - y}$

$= \dfrac{(x + y)(x - y)}{(x + 7)(x - 7)} \cdot \dfrac{x + 7}{x - y}$ *Factor*

$= \dfrac{(x + y)\cancel{(x - y)}^1}{_1\cancel{(x + 7)}(x - 7)} \cdot \dfrac{\cancel{x + 7}^1}{\cancel{x - y}_1}$ *Cancel*

$= \dfrac{x + y}{x - 7}$ *Multiply*

4) $\dfrac{x^2 - 1}{4} \cdot \dfrac{12}{x - 1}$

$= \dfrac{(x + 1)(x - 1)}{4} \cdot \dfrac{12}{x - 1}$ *Factor*

$= \dfrac{(x + 1)\cancel{(x - 1)}^1}{\cancel{4}_1} \cdot \dfrac{^3\cancel{12}}{\cancel{x - 1}_1}$ *Cancel*

$= \dfrac{3(x + 1)}{1}$

$= 3(x + 1)$

Chapter 7: **RATIONAL EXPRESSIONS**

EXERCISES

Simplify each of the following.

1) $\dfrac{x}{3} \cdot \dfrac{6}{x}$

2) $\dfrac{x}{3} \cdot \dfrac{6}{2x^3}$

3) $\dfrac{9x^2}{7y} \cdot \dfrac{14xy}{3x}$

4) $\dfrac{3}{7} \cdot \dfrac{7x-7}{6}$

5) $\dfrac{7}{x^2-4} \cdot \dfrac{2x+4}{21}$

6) $\dfrac{a}{x^2-4} \cdot (2x+4)$

7) $\dfrac{1}{x+2} \cdot \dfrac{x^2+2x}{4}$

8) $\dfrac{x^2-1}{4} \cdot \dfrac{8}{x-1}$

9) $\dfrac{6}{x^2-4} \cdot \dfrac{x-2}{18}$

10) $\dfrac{a+3}{a-5} \cdot \dfrac{2a-10}{3a+9}$

11) $\dfrac{c}{a+b} \cdot \dfrac{5a+5b}{2c^2+2c}$

12) $\dfrac{x^2+6x+9}{8} \cdot \dfrac{4x+8}{x^2+5x+6}$

13) $\dfrac{x^2-1}{14} \cdot \dfrac{7}{x+1}$

14) $\dfrac{3x+9}{15x} \cdot \dfrac{x^2}{x^2-9}$

15) $\dfrac{3x}{x+4} \cdot 3x+12$

16) $\dfrac{x^2-1}{x} \cdot \dfrac{4x^2}{x+1}$

17) $\dfrac{3x^2}{x^2-9} \cdot \dfrac{x+3}{x}$

18) $\dfrac{(x+5)^2}{25} \cdot \dfrac{5}{x+5}$

REVIEW

Simplify each algebraic fraction.

1) $\dfrac{4x}{12xy}$

2) $\dfrac{8a^2x}{4ax^2}$

3) $\dfrac{12a^5b^4}{4a^3b}$

4) $\dfrac{-50x^2y^4}{10xy^2}$

7-6 Dividing Algebraic Fractions

Dividing algebraic fractions is very similar to dividing arithmetic fractions. Simply **invert** the **second** fraction and then multiply the two algebraic fractions. After inverting the second fraction, use the exact procedure that you used in multiplying algebraic fractions. Remember the following.

Helpful Hints
- **First**, invert the second fraction.
- **Second**, factor the numerators and denominators of each fraction if possible.
- **Third**, carefully cancel out any common factors.
- **Fourth**, to get the final answer multiply the remaining numerators and the remaining denominators.
- It may be good to review the laws of exponents.

EXAMPLES

Divide each of the following.

1) $\dfrac{2}{5x} \div \dfrac{2x}{4}$

$= \dfrac{2}{5x} \cdot \dfrac{4}{2x}$ *Invert*

$= \dfrac{\overset{1}{2}}{5x} \cdot \dfrac{4}{\underset{1}{2}x}$ *Cancel*

$= \dfrac{4}{5x^2}$ *Multiply*

2) $\dfrac{5x^2}{9} \div \dfrac{4x}{3}$

$= \dfrac{5x^2}{9} \cdot \dfrac{3}{4x}$ *Invert*

$= \dfrac{5x^2}{\underset{3}{9}} \cdot \dfrac{\overset{1}{3}}{4x}$ *Cancel*

$= \dfrac{5x^2}{12x}$ *Multiply*

$= \dfrac{5x^{2-1}}{12}$

$= \dfrac{5x}{12}$

3) $\dfrac{9}{y^2 - 9} \div \dfrac{3}{y - 3}$

$= \dfrac{9}{y^2 - 9} \cdot \dfrac{y - 3}{3}$ *Invert*

$= \dfrac{9}{(y + 3)(y - 3)} \cdot \dfrac{y - 3}{3}$ *Factor*

$= \dfrac{\overset{3}{9}}{(y + 3)(y - 3)_1} \cdot \dfrac{\overset{1}{y - 3}}{\underset{1}{3}}$ *Cancel*

$= \dfrac{3}{y + 3}$ *Multiply*

4) $\dfrac{28x^2y^3}{5} \div 7x^2y$

$= \dfrac{28x^2y^3}{5} \cdot \dfrac{1}{7x^2y}$ *Invert*

$= \dfrac{\overset{4}{28}\overset{1}{x^2}y^2}{5} \cdot \dfrac{1}{\underset{1}{7}x^2_1y_1}$ *Cancel*

$= \dfrac{4y^2}{5}$ *Multiply*

7-6 Dividing Algebraic Fractions

Divide each of the following.

1) $\dfrac{x}{4} \div \dfrac{x}{5}$

2) $\dfrac{x^2}{4x} \div \dfrac{x}{16}$

3) $\dfrac{xy^2}{x^2y} \div \dfrac{x}{y^2}$

4) $\dfrac{3x^2}{5y^2} \div \dfrac{6x^3}{15y}$

5) $\dfrac{6}{x^2y} \div \dfrac{x}{x^3y^2}$

6) $\dfrac{2n-2}{4} \div \dfrac{2}{9}$

7) $\dfrac{4x+4}{5} \div \dfrac{2}{3}$

8) $\dfrac{12}{(x^2-9)} \div \dfrac{3}{x-3}$

9) $\dfrac{x-y}{4} \div \dfrac{x-y}{20}$

10) $\dfrac{x^2-y^2}{3} \div \dfrac{x-y}{15}$

11) $\dfrac{9}{x^2-1} \div \dfrac{3}{x+1}$

12) $\dfrac{x^2-y^2}{10} \div \dfrac{3x-3y}{2}$

13) $\dfrac{x+1}{2y-2} \div \dfrac{x+1}{y-1}$

14) $\dfrac{3x^2}{x^2-4} \div \dfrac{6x}{x+2}$

15) $\dfrac{2x-3}{3x-2} \div \dfrac{4x-6}{9x-6}$

16) $\dfrac{2x-1}{x^3+3x} \div \dfrac{x+1}{x^2+3}$

17) $\dfrac{x^2-y^2}{2xy} \div \dfrac{x-y}{4x^2}$

18) $\dfrac{x}{y^2} \div \dfrac{x-y}{y^3}$

REVIEW

Solve each of the following.

1) $3x + 4x - 7 = 2x + 3$

2) $2x - x + 10 = 5x - 6$

3) $7x + 2 = 4x + 4$

4) $5x + 10 = 9x - 6$

7-7 Adding and Subtracting Algebraic Fractions with Like Denominators

INTRODUCTION

Adding and subtracting algebraic fractions is very similar to how you added and subtracted fractions back in elementary school. The only difference is that the numerators and denominators contain variables. Remember the following when completing the addition or subtraction.

Helpful Hints

- **First**, write a fraction whose numerator is the sum or difference of the numerators you are working with. The denominator will be the one given in the problem.

- **Second**, reduce the fraction to its lowest terms. Sometimes it is necessary to factor either the numerator or denominator, or both.

- Remember to be careful with signs when **subtracting**. These steps can prevent mistakes when parentheses are involved.

  First, remove the parentheses.

  Second, get rid of the minus sign (–) and change the sign of each term.

  Third, collect the like terms and add the coefficients.

  Example: $5x - (5 - 2x) = 5x - 5 + 2x = 7x - 5$

- When adding, remove the parentheses and the plus sign (+). Do not change the signs of the terms.

EXAMPLES

Add or subtract each of the following.

1) $\dfrac{3}{2x} + \dfrac{4}{2x}$

$= \dfrac{3+4}{2x}$ *Add*

$= \dfrac{7}{2x}$

2) $\dfrac{3x}{x-y} - \dfrac{3y}{x-y}$

$= \dfrac{3x-3y}{x-y}$ *Subtract*

$= \dfrac{3(\cancel{x-y})^1}{(\cancel{x-y})_1}$ *Factor and Cancel*

$= 3$

3) $\dfrac{5x+6}{x-3} - \dfrac{2x-4}{x-3}$

$= \dfrac{(5x+6)-(2x-4)}{x-3}$ *Subtract*

$= \dfrac{5x+6-2x+4}{x-3}$ *Be careful with signs*

$= \dfrac{3x+10}{x-3}$

4) $\dfrac{5x}{4} - \dfrac{3-2x}{4}$

$= \dfrac{5x-(3-2x)}{4}$ *Subtract*

$= \dfrac{5x-3+2x}{4}$ *Be careful with signs*

$= \dfrac{7x-3}{4}$

       Chapter 7: **RATIONAL EXPRESSIONS**

7-7 Adding and Subtracting Algebraic Fractions with Like Denominators

Add or subtract each of the following.

1) $\dfrac{8}{3x} + \dfrac{3}{3x}$

2) $\dfrac{5x}{9} - \dfrac{2y}{9}$

3) $\dfrac{3}{2x} + \dfrac{1}{2x}$

4) $\dfrac{5x}{x+1} + \dfrac{2x}{x+1}$

5) $\dfrac{3x}{11} + \dfrac{x+5}{11}$

6) $\dfrac{r+s}{3rs} + \dfrac{2r-s}{3rs}$

7) $\dfrac{3x+2y}{4xy} + \dfrac{x+2y}{4xy}$

8) $\dfrac{2x+1}{7x} - \dfrac{4x-3}{7x}$

9) $\dfrac{3m+2}{m+3} - \dfrac{m-4}{m+3}$

10) $\dfrac{2x+1}{2} + \dfrac{3x+6}{2}$

11) $\dfrac{8x-4}{2x+6} - \dfrac{4x-6}{2x+6}$

12) $\dfrac{x}{x^2-4} - \dfrac{2}{x^2-4}$

13) $\dfrac{8x-3}{5} - \dfrac{2x+7}{5}$

14) $\dfrac{3x+4}{2x-1} - \dfrac{x+5}{2x-1}$

15) $\dfrac{7x-8}{x^2-4} - \dfrac{6x-10}{x^2-4}$

16) $\dfrac{x^2-4x}{x-2} + \dfrac{x+2}{x-2}$

REVIEW

Simplify each of the following.

1) $\dfrac{4n-8}{8}$

2) $\dfrac{x-1}{2x-2}$

3) $\dfrac{(m-4)^2}{3m-12}$

4) $\dfrac{x^2-4x+4}{x^2-4}$

7-8 Adding and Subtracting Algebraic Fractions with Unlike Denominators

When adding or subtracting algebraic fractions with unlike denominators, it is necessary to find a lowest common denominator. Once the fractions are renamed and have the same denominators, it is easy to add or subtract them. Again, this is very similar to what you did with arithmetic fractions back in elementary school. Remember the following steps when completing your addition or subtraction.

Helpful Hints
- **First**, find the lowest common denominator (LCD). Basically, the LCD is the smallest expression that is divisible by each of the given denominators.

  Example: The LCD of $4x^2y$ and $2xy^2$ is $4x^2y^2$.

- **Second**, rename each fraction with the new denominator.

- **Third**, complete the addition or subtraction. When subtracting, be care of the signs.

- **Fourth**, reduce your answer to lowest terms.

EXAMPLES

Add or subtract each of the following. Reduce answers to lowest terms.

1) $\dfrac{5}{2x} + \dfrac{1}{2}$ *The LCD is 2x*

$= \dfrac{5}{2x} + \dfrac{x}{x} \cdot \dfrac{1}{2}$ *Rename*

$= \dfrac{5}{2x} + \dfrac{x}{2x}$

$= \dfrac{5+x}{2x}$

2) $\dfrac{3}{xy} + \dfrac{5}{2x}$ *The LCD is 2xy*

$= \dfrac{2}{2} \cdot \dfrac{3}{xy} + \dfrac{y}{y} \cdot \dfrac{5}{2x}$ *Rename*

$= \dfrac{6}{2xy} + \dfrac{5y}{2xy}$

$= \dfrac{6+5y}{2xy}$

3) $\dfrac{x+3}{3} + \dfrac{x+2}{5}$ *The LCD is 15*

$= \dfrac{5}{5} \cdot \dfrac{(x+3)}{3} + \dfrac{3}{3} \cdot \dfrac{(x+2)}{5}$ *Rename using distributive property*

$= \dfrac{5x+15}{15} + \dfrac{3x+6}{15}$

$= \dfrac{5x+15+3x+6}{15}$ *Collect like terms*

$= \dfrac{8x+21}{15}$

4) $\dfrac{2x-1}{4} - \dfrac{x+2}{8}$ *The LCD is 8*

$= \dfrac{2}{2} \cdot \dfrac{(2x-1)}{4} - \dfrac{x+2}{8}$ *Rename*

$= \dfrac{(4x-2)}{8} - \dfrac{(x+2)}{8}$

$= \dfrac{(4x-2)-(x+2)}{8}$

$= \dfrac{4x-2-x-2}{8}$ *Collect like terms*

$= \dfrac{3x-4}{8}$

7-8 Adding and Subtracting Algebraic Fractions with Unlike Denominators

Add or subtract each of the following. Reduce answers to lowest terms.

1) $\dfrac{5x}{2} + \dfrac{x}{4}$

2) $\dfrac{5x}{3} - \dfrac{x}{6}$

3) $\dfrac{7}{2x} - \dfrac{1}{4x}$

4) $\dfrac{7}{x} + \dfrac{3}{4x}$

5) $\dfrac{4}{3x} + \dfrac{1}{6xy}$

6) $\dfrac{3x}{8} - \dfrac{x}{12}$

7) $\dfrac{5}{3x} + \dfrac{1}{4x}$

8) $\dfrac{5x}{4} - \dfrac{2x}{3} + \dfrac{x}{6}$

9) $\dfrac{3x}{10} + \dfrac{7x}{5}$

10) $\dfrac{3x - 2}{3} + \dfrac{2x - 1}{6}$

11) $\dfrac{4x + 3}{4} + \dfrac{x - 1}{3}$

12) $\dfrac{2x + 3}{5} - \dfrac{x - 1}{2}$

13) $\dfrac{5x^2}{3} + \dfrac{3x^2}{4}$

14) $\dfrac{8}{m^2n} + \dfrac{5}{mn^2}$

15) $\dfrac{x + 2}{3} + \dfrac{x - 3}{7}$

16) $\dfrac{3x}{x^2 - 9} + \dfrac{3}{x - 3}$

Cross multiply to solve each proportion.

1) $\dfrac{4}{5} = \dfrac{2x}{25}$

2) $\dfrac{x + 1}{3} = \dfrac{x}{2}$

3) $\dfrac{x + 3}{5} = \dfrac{x + 1}{4}$

4) $\dfrac{x - 3}{x} = \dfrac{x + 3}{x + 8}$

7-9 Adding and Subtracting Algebraic Fractions with Binomial Denominators

INTRODUCTION

Sometimes it is necessary to add or subtract fractions that have binomial denominators. Just remember to use the same techniques that were used in simpler problems. There will often be quite a few steps, so it is necessary to be very careful. Remember the following when completing your work.

Helpful Hints
- **First**, find the lowest common denominator. Sometimes a denominator will have to be factored first.
- **Second**, rename each fraction with the new denominator.
- **Third**, complete the addition or subtraction.
- **Fourth**, reduce your answer to lowest terms.
- Be careful with signs.

EXAMPLES

Add or subtract each of the following.

1) $\dfrac{6}{x+3} + \dfrac{3}{x-1}$ *The LCD is (x + 3) (x – 1)*

$= \dfrac{6}{(x+3)} \cdot \dfrac{(x-1)}{(x-1)} + \dfrac{3}{(x-1)} \cdot \dfrac{(x+3)}{(x+3)}$ *Rename and multiply using the distributive property*

$= \dfrac{6x-6}{(x+3)\,(x-1)} + \dfrac{3x+9}{(x+3)\,(x-1)}$

$= \dfrac{6x-6+3x+9}{(x+3)\,(x-1)}$ *Collect like terms*

$= \dfrac{9x+3}{(x+3)\,(x-1)}$

2) $\dfrac{x}{x^2-9} - \dfrac{1}{x+3}$ *First, factor $x^2 – 9$*

$= \dfrac{x}{(x+3)\,(x-3)} - \dfrac{1}{x+3}$ *The LCD is (x + 3)(x – 3)*

$= \dfrac{x}{(x+3)\,(x-3)} - \dfrac{1}{(x+3)} \cdot \dfrac{(x-3)}{(x-3)}$ *Rename and multiply*

$= \dfrac{x}{(x+3)\,(x-3)} - \dfrac{x-3}{(x+3)\,(x-3)}$

$= \dfrac{x-(x-3)}{(x+3)(x-3)}$

$= \dfrac{x-x+3}{(x+3)(x-3)}$

$= \dfrac{3}{(x+3)(x-3)} = \dfrac{3}{x^2-9}$

7-9 Adding and Subtracting Algebraic Fractions with Binomial Denominators

Add or subtract each of the following.

1) $\dfrac{x}{x+3} + \dfrac{x}{x+2}$

2) $\dfrac{4}{x+3} - \dfrac{x}{x-2}$

3) $\dfrac{6}{x+4} + \dfrac{3}{x-1}$

4) $\dfrac{x}{x-3} - \dfrac{5}{x+4}$

5) $\dfrac{5}{x^2-9} + \dfrac{3}{x-3}$

6) $\dfrac{6}{x^2-16} - \dfrac{5}{x+4}$

7) $\dfrac{5}{x-3} + \dfrac{7}{2x-6}$

8) $\dfrac{3}{x+6} - \dfrac{5}{x^2-36}$

9) $\dfrac{x}{x+4} + \dfrac{x}{x+2}$

10) $\dfrac{6}{x+1} - \dfrac{8}{x-2}$

11) $\dfrac{5}{x^2-9} + \dfrac{1}{x-3}$

12) $\dfrac{7}{x-1} + \dfrac{7}{2x+3}$

13) $\dfrac{9}{x+1} - \dfrac{3}{4x+4}$

14) $\dfrac{10}{3x-6} + \dfrac{3}{2x-4}$

REVIEW

Multiply each of the following.

1) $\dfrac{a}{b} \cdot \dfrac{2b}{3a}$

2) $\dfrac{3a^2}{4} \cdot \dfrac{20b}{12b^3}$

3) $\dfrac{x^2-y^2}{x^2-9} \cdot \dfrac{x+3}{x-y}$

4) $\dfrac{x^2-1}{4} \cdot \dfrac{12}{x-1}$

7-10 Solving Equations Involving Algebraic Fractions

Equations will often contain algebraic fractions. There are basically **two** ways to solve these equations. The **first** way is to clear out all the fractions and then solve the equation. The **second** way can be used when there are two fractions, one on each side of the equal sign. In this case we have a proportion. Remember, to solve a proportion, simply cross-multiply. Remember the following when solving equations involving algebraic fractions.

Helpful Hints
- If the equation contains fractions with the **same denominator**, clear out the fractions by multiplying each term by the given denominator. Then solve the equation.

- If the equation contains fractions with **different denominators**, clear out the fractions by multiplying each term by the lowest common denominator. Then solve the equation.

- If the equation represents a **proportion**, simply cross-multiply and solve.

EXAMPLES

Solve each of the following.

1) $\dfrac{3x}{4} = \dfrac{x}{4} + 10$ *The denominator is 4*

$4 \cdot \dfrac{3x}{4} = 4 \cdot (\dfrac{x}{4} + 10)$ *Multiply*

$3x = x + 40$ *Solve for x*

$2x = 40$

$x = 20$

2) $\dfrac{2x}{3} = \dfrac{x}{4} + 10$ *The LCD is 12*

$^{4}\cancel{12} \cdot \dfrac{2x}{\cancel{3}_1} = 12 \,(\dfrac{x}{4} + 10)$ *Multiply*

$8x = 3x + 120$ *Solve for x*

$5x = 120$

$x = 24$

3) $\dfrac{x+2}{3} + \dfrac{x-1}{6} = 5$ *The LCD is 6*

$6 \cdot \dfrac{x+2}{3} + 6 \cdot \dfrac{x-1}{6} = 5 \cdot 6$ *Multiply*

$2(x+2) + x - 1 = 30$ *Solve for x*

$2x + 4 + x - 1 = 30$

$3x + 3 = 30$

$3x = 27$

$x = 9$

4) $\dfrac{6x-9}{5} = \dfrac{2x+1}{3}$ *This is a proportion*

$3(6x - 9) = 5(2x + 1)$ *Cross multiply*

$18x - 27 = 10x + 5$ *Solve for x*

$8x - 27 = 5$

$8x = 32$

$x = 4$

7-10 Solving Equations Involving Algebraic Fractions

Solve each of the following.

1) $\frac{4x}{3} = \frac{x}{3} + 6$

2) $\frac{x}{3} + \frac{x}{6} = 1$

3) $\frac{x}{2} = \frac{x}{4} + 2$

4) $\frac{y + 2}{4} = \frac{5}{2}$

5) $\frac{x}{3} + \frac{x}{2} = 40$

6) $\frac{3x}{4} - \frac{2x}{3} = 1$

7) $\frac{x}{3} - \frac{x}{6} = 2$

8) $\frac{2x}{5} - \frac{x}{4} = 3$

9) $\frac{x + 1}{2} + \frac{2x - 3}{3} = 10$

10) $\frac{x - 1}{4} - \frac{2x - 3}{4} = 5$

11) $\frac{x}{3} + \frac{x}{7} = 10$

12) $\frac{2x}{3} = \frac{3x + 9}{4}$

13) $\frac{2x}{5} - 4 = \frac{2x}{3}$

14) $\frac{x + 2}{3} - \frac{x - 2}{5} = 2$

15) $\frac{x}{8} - \frac{x}{10} = 3$

16) $\frac{3x - 4}{3} = \frac{2x + 4}{6}$

REVIEW

Divide each of the following.

1) $\frac{2}{3m} \div \frac{4}{m}$

2) $\frac{15r^2}{12s^2} \div \frac{5r}{24s^3}$

3) $\frac{6}{y^2 - 9} \div \frac{3}{y - 3}$

4) $\frac{2x - 2}{4} \div \frac{2}{9}$

Chapter 7 Review: Rational Expressions

Simplify each of the following. For questions 4-6, remember the -1 factor.

1) $\dfrac{56x^3y}{8x}$

2) $\dfrac{12xy + 4y}{4y}$

3) $\dfrac{x - y}{y - x}$

4) $\dfrac{x - y}{4y - 4x}$

5) $\dfrac{x^2 - y^2}{x + y}$

6) $\dfrac{x^2 - 16}{4 - x}$

7) Solve the proportion. $\dfrac{x + 3}{x - 3} = \dfrac{6}{5}$

Add, subtract, multiply or divide each of the following.

8) $\dfrac{x^2 - y^2}{x} \cdot \dfrac{4x}{x + y}$

9) $\dfrac{3}{5} \cdot \dfrac{5m - 5}{9}$

10) $\dfrac{12}{x^2 - 16} \div \dfrac{3}{x - 4}$

11) Solve. $\dfrac{2}{5} + \dfrac{3}{10} = \dfrac{7}{n}$

12) $\dfrac{2}{3x} - \dfrac{5}{2x}$

13) $\dfrac{x}{4} + \dfrac{y}{8}$

14) $\dfrac{x - 4}{3} + \dfrac{x}{4}$

15) $\dfrac{x + y}{3} - \dfrac{x - y}{4}$

Solve each of the following.

16) $\dfrac{1}{x} + \dfrac{1}{2} = \dfrac{3}{x}$

17) $\dfrac{5}{x + 3} = \dfrac{4}{x + 2}$

Don't Forget the Resources in the Back of the Book

There are several useful resources in the back of this book. If you have not taken the time to look through them, do so. When necessary, put them to good use.

Here is a list of the useful resources that are available to you.

- Glossary

- Important Formulas

- Important Symbols

- Multiplication Table

- Commonly Used Prime Numbers

- Squares and Square Roots

- Fraction/Decimal Equivalents

- Solutions

8-1 Simplifying Radical Expressions

INTRODUCTION

Radical expressions and **square roots** mean basically the same thing. You learned the basics of square roots in chapter one. The symbol for square root is $\sqrt{}$. This symbol is also called a **radical sign**. The expression $\sqrt{64}$ is read, "the square root of 64." The answer to this expression is the number which, when multiplied by itself, is equal to 64. That number is 8. The number 64 is an example of a **perfect square** since its square root is a whole number.

An **irrational number** is a number such as $\sqrt{5}$ that cannot be expressed as the ratio of two integers. When working with square roots, irrational numbers are often left as a radical. Many radical expressions will contain **numbers** and **variables**. Remember the following facts when simplifying radical expressions.

Helpful Hints
- $\sqrt{a^2} = a$
- $\sqrt{a \times b} = \sqrt{a} \times \sqrt{b}$
- If the answer is not a perfect square, leave your answer as the product of a perfect square and a radical. The radical will be the last part of the answer.
- There is no exact order that has to be followed. With practice you can get really good.

EXAMPLES

1) $\sqrt{3600}$
   $= \sqrt{36 \cdot 100}$
   $= \sqrt{36} \cdot \sqrt{100}$
   $= 6 \times 10 = 60$

2) $\sqrt{27}$
   $= \sqrt{9 \cdot 3}$
   $= \sqrt{9} \cdot \sqrt{3}$
   $= 3\sqrt{3}$

3) $\sqrt{36x^2 y}$
   $= \sqrt{36} \cdot \sqrt{x^2} \cdot \sqrt{y}$
   $= 6x\sqrt{y}$

4) $\sqrt{125x^3}$
   $= \sqrt{25 \cdot 5 \cdot x^2 \cdot x}$
   $= \sqrt{25} \cdot \sqrt{5} \cdot \sqrt{x^2} \cdot \sqrt{x}$
   $= 5\sqrt{5} \times \sqrt{x}$
   $= 5x\sqrt{5x}$

5) $\sqrt{x^3 y^3}$
   $= \sqrt{x^2 \cdot x \cdot y^2 \cdot y}$
   $= \sqrt{x^2} \cdot \sqrt{x} \cdot \sqrt{y^2} \cdot \sqrt{y}$
   $= x\sqrt{x} \cdot y\sqrt{y}$
   $= xy\sqrt{x} \cdot \sqrt{y}$
   $= xy\sqrt{xy}$

6) $3\sqrt{mn^2}$
   $= 3\sqrt{m} \cdot \sqrt{n^2}$
   $= 3\sqrt{m}\ n$
   $= 3n\sqrt{m}$

8-1 Simplifying Radical Expressions

Simplify each radical expression.

1) $\sqrt{400}$

2) $\sqrt{45}$

3) $\sqrt{99}$

4) $\sqrt{300}$

5) $\sqrt{25x^2}$

6) $\sqrt{12x^2}$

7) $\sqrt{45x^2 y^2}$

8) $\sqrt{x^2 y^3}$

9) $\sqrt{49x^3}$

10) $\sqrt{16x^2 y^3}$

11) $\sqrt{9x^3 y^3}$

12) $\sqrt{50x^2 y^2 z^2}$

13) $\sqrt{16y^3 z^3}$

14) $5\sqrt{18x^4 y^3}$

15) $\sqrt{72x^4 y^6}$

16) $3\sqrt{50m^2 n^2 p}$

17) $\sqrt{x^5 y^5}$

18) $\sqrt{18x^8}$

REVIEW

Add or subtract each of the following.

1) $\frac{3}{2m} + \frac{1}{2}$

2) $\frac{4}{3x} + \frac{1}{4x}$

3) $\frac{1}{a} - \frac{3}{2ab}$

4) $\frac{3}{2a} - \frac{1}{4a}$

8-2 Solving Equations Involving Radicals

INTRODUCTION

Sometimes you will use square roots when solving equations. Just as in other equations, it is necessary to isolate the variable and solve for its value. There will often be two solutions to an equation with a radical. For example:

$$\sqrt{36} = 6 \text{ or } -6, \text{ since } 6 \times 6 = 36 \text{ and } -6 \times -6 = 36$$

Remember the following when solving equations involving radicals.

Helpful Hints
- Isolate the radical on the left side of the equal sign.
- Solve for the variable.
- Sometimes it will be necessary to square both sides of the equation.
- Sometimes it will be necessary to take the square root of both sides of the equation.
- There will often be two solutions.

EXAMPLES

Solve each of the following. Check your answer by substituting back into the original equation.

1) $x^2 = 49$ *Take the square root of both sides*

$x = \pm 7$

$x = 7 \ or \ x = -7$

2) $\dfrac{n}{16} = \dfrac{4}{n}$ *Cross multiply*

$n^2 = 64$ *Take the square root of both sides*

$n = \pm 8$

3) $x^2 - 5 = 11$

$x^2 - 5 + 5 = 11 + 5$ *Add 5 to both sides*

$x^2 = 16$ *Take the square root of both sides*

$x = \pm 4$

4) $\sqrt{3x} = 12$ *Square both sides*

$(\sqrt{3x})^2 = 12^2$

$3x = 144$

$\dfrac{3x}{3} = \dfrac{144}{3}$ *Divide both sides by 3*

$x = 48$

5) $\sqrt{2x} + 3 = 7$

$\sqrt{2x} + 3 - 3 = 7 - 3$ *Isolate the radical*

$\sqrt{2x} = 4$

$(\sqrt{2x})^2 = 4^2$ *Square both sides*

$2x = 16$ *Solve for x*

$x = 8$

6) $\sqrt{2x + 1} - 1 = 4$

$\sqrt{2x + 1} - 1 + 1 = 4 + 1$ *Isolate the radical*

$\sqrt{2x + 1} = 5$

$(\sqrt{2x + 1})^2 = 5^2$ *Square both sides*

$2x + 1 = 25$

$2x + 1 - 1 = 25 - 1$ *Solve for x*

$2x = 24$

$x = 12$

EXERCISES

Solve each of the following. Check your answer by substituting back into the original equation.

1) $x^2 = 81$

2) $\frac{25}{x} = \frac{x}{4}$

3) $\frac{n}{2} = \frac{8}{n}$

4) $x^2 - 5 = 11$

5) $x^2 + 5 = 21$

6) $x^2 + 7 = 52$

7) $\sqrt{x} = 2$

8) $\sqrt{x + 4} = 3$

9) $2\sqrt{x} - 12 = 0$

10) $\sqrt{2x - 1} = 7$

11) $\sqrt{5x - 1} - 3 = 0$

12) $\sqrt{2x + 1} - 1 = 4$

13) $\sqrt{3x + 3} = 6$

14) $\sqrt{x - 2} + 4 = 6$

15) $2\sqrt{5x} = 20$

16) $\sqrt{4x - 3} = \sqrt{3x + 4}$

17) $4\sqrt{3x + 3} = 24$

18) $8\sqrt{x + 3} = 64$

REVIEW

Add or subtract each of the following. Reduce answers to lowest terms.

1) $\frac{7}{3n} - \frac{4}{9m}$

2) $\frac{x + 2}{3} + \frac{x + 1}{5}$

3) $\frac{3x}{x - 1} - \frac{3}{x - 1}$

4) $\frac{3}{3x} + \frac{2}{5x}$

8-3 Adding and Subtracting Radical Expressions

Radicals can be **added** or **subtracted** if they have **like terms**. Simply add or subtract the coefficients.

Examples: $3\sqrt{2} + 4\sqrt{2} = (3+4)\sqrt{2} = 7\sqrt{2}$

$4\sqrt{3} + 5\sqrt{2}$ has **unlike** terms and cannot be added.

However, unlike radicals can sometimes be simplified so that they will have like terms. Then it is possible to add or subtract. You will see this in the examples below. Remember the following when adding or subtracting radicals.

Helpful Hints
- To add or subtract **like radicals**, just add or subtract the coefficients.

- To add or subtract **unlike radicals**, sometimes it is possible to simplify them so that they have like terms. Then you can add or subtract.

EXAMPLES

Add or subtract each of the following.

1) $3\sqrt{5} + 4\sqrt{5}$

$= (3+4)\sqrt{5}$

$= 7\sqrt{5}$

2) $9\sqrt{3} - \sqrt{3}$

$= (9-1)\sqrt{3}$

$= 8\sqrt{3}$

3) $\sqrt{50} + \sqrt{2}$

$= \sqrt{25 \cdot 2} + \sqrt{2}$ *Simplify*

$= \sqrt{25} \cdot \sqrt{2} + \sqrt{2}$ *Simplify*

$= 5\sqrt{2} + \sqrt{2}$ *Add like terms*

$= 6\sqrt{2}$

4) $5\sqrt{3} + 3\sqrt{12}$

$= 5\sqrt{3} + 3\sqrt{4} \cdot \sqrt{3}$ *Simplify*

$= 5\sqrt{3} + 3 \cdot 2 \cdot \sqrt{3}$

$= 5\sqrt{3} + 6\sqrt{3}$ *Add like terms*

$= 11\sqrt{3}$

5) $\sqrt{12} - 4\sqrt{3} + \sqrt{6}$

$= \sqrt{4} \cdot \sqrt{3} - 4\sqrt{3} + \sqrt{6}$ *Simplify*

$= 2\sqrt{3} - 4\sqrt{3} + \sqrt{6}$ *Subtract like terms*

$= -2\sqrt{3} + \sqrt{6}$

6) $3\sqrt{32} - 6\sqrt{8}$

$= 3\sqrt{4} \cdot \sqrt{8} - 6\sqrt{8}$ *Simplify*

$= 3 \cdot 2\sqrt{8} - 6\sqrt{8}$

$= 6\sqrt{8} - 6\sqrt{8}$ *Subtract like terms*

$= 0$

8-3 Adding and Subtracting Radical Expressions

Add or subtract each of the following. Express in simplest form.

1) $9\sqrt{2} + 3\sqrt{2}$

2) $7\sqrt{3} - 2\sqrt{3}$

3) $5\sqrt{3} + \sqrt{3} - 2\sqrt{3}$

4) $4\sqrt{7} - \sqrt{7} - 5\sqrt{7}$

5) $\sqrt{27} + \sqrt{75}$

6) $\sqrt{18} + \sqrt{2}$

7) $6\sqrt{3} - 4\sqrt{27}$

8) $3\sqrt{12} + 4\sqrt{3} + \sqrt{5}$

9) $\sqrt{45} + 2\sqrt{5} - 4\sqrt{7}$

10) $\sqrt{2} + \sqrt{8} - 3\sqrt{8}$

11) $5\sqrt{3} + \sqrt{48}$

12) $\sqrt{27} - \sqrt{3} - \sqrt{12}$

13) $\sqrt{72} - \sqrt{50}$

14) $3\sqrt{2} + 2\sqrt{32}$

15) $3\sqrt{50} - 5\sqrt{18}$

16) $\sqrt{3x^2} + \sqrt{12x^2}$

17) $\sqrt{81x} - \sqrt{25x}$

18) $\sqrt{80} - \sqrt{45}$

REVIEW

Add or subtract each of the following.

1) $\dfrac{5}{x} + \dfrac{3}{x+1}$

2) $\dfrac{4}{x+1} - \dfrac{3}{x-1}$

8-4 Multiplying Radical Expressions

INTRODUCTION

It is easy to multiply radical expressions. Remember the following.

Helpful Hints

• The product of the square roots is equal to the square root of the product.

$$\sqrt{a} \cdot \sqrt{b} = \sqrt{ab}$$

• To multiply a binomial radical by a monomial radical it is necessary to use the distributive property.

• To multiply two binomial radicals use the same technique that would be used for any two binomials. The rules are the same. Remember the **FOIL technique**.

EXAMPLES

Multiply and simplify each of the following.

1) $\sqrt{10} \cdot \sqrt{2}$

$= \sqrt{10 \cdot 2}$

$= \sqrt{20}$

$= \sqrt{4 \cdot 5}$

$= \sqrt{4} \cdot \sqrt{5}$

$= 2\sqrt{5}$

2) $3\sqrt{6} \cdot 5\sqrt{2}$

$= 3 \cdot 5 \cdot \sqrt{6} \cdot \sqrt{2}$

$= 15\sqrt{12}$

$= 15\sqrt{4 \cdot 3}$

$= 15\sqrt{4} \cdot \sqrt{3}$

$= 15 \cdot 2 \cdot \sqrt{3}$

$= 30\sqrt{3}$

3) $\sqrt{3}\,(7 - \sqrt{3})$

$= \sqrt{3}\,(7 - \sqrt{3})$ *Use the distributive property*

$= 7\sqrt{3} - \sqrt{3} \cdot \sqrt{3}$

$= 7\sqrt{3} - 3$

4) $(4 + \sqrt{2})(2 - \sqrt{2})$

$= 4(2) - 4 \cdot \sqrt{2} + 2\sqrt{2} - \sqrt{2} \cdot \sqrt{2}$ *Use the foil method*

$= 8 - 2\sqrt{2} - 2$ *Simplify*

$= 6 - 2\sqrt{2}$

Chapter 8: **RADICAL EXPRESSIONS AND GEOMETRY**

8-4 Multiplying Radical Expressions

Multiply and simplify each of the following.

1) $\sqrt{32} \cdot \sqrt{2}$

2) $\sqrt{14} \cdot \sqrt{2}$

3) $\sqrt{21} \cdot \sqrt{3}$

4) $3\sqrt{6} \cdot 5\sqrt{2}$

5) $\sqrt{3} \cdot 2\sqrt{4} \cdot 3\sqrt{3}$

6) $5\sqrt{8} \cdot 7\sqrt{3}$

7) $2\sqrt{n} \cdot \sqrt{4n}$

8) $2\sqrt{2} \cdot \sqrt{20}$

9) $3(\sqrt{3} - 3)$

10) $2\sqrt{3}(2 - \sqrt{6})$

11) $(\sqrt{2} + \sqrt{3})(\sqrt{2} - \sqrt{3})$

12) $(3 + \sqrt{5})(4 - 2\sqrt{5})$

13) $\sqrt{3} \cdot \sqrt{75}$

14) $\sqrt{6}(\sqrt{6} - 1)$

15) $\sqrt{6}(2\sqrt{3} - 4\sqrt{2})$

16) $(2 + \sqrt{3})(2 - \sqrt{3})$

17) $\sqrt{2}(\sqrt{8} - 4)$

18) $(\sqrt{3} - 2)^2$

REVIEW

Solve each of the following.

1) $\frac{x}{6} + \frac{x}{7} = 13$

2) $\frac{x}{3} - \frac{x}{5} = 4$

3) $\frac{2}{5} + \frac{3}{10} = \frac{7}{n}$

4) $\frac{3n}{7} - \frac{n}{3} = 4$

8-5 Dividing Radical Expressions

Dividing monomials that contain radicals is a simple process. It is important to know the following rule.

$$\frac{\sqrt{a}}{\sqrt{b}} = \sqrt{\frac{a}{b}}$$

It is often necessary to divide the coefficients and radicals separately, and then simplify if possible. Follow the following steps when completing your work.

Helpful Hints
- **First**, if there are coefficients, divide them.
- **Second**, divide the radicals.
- **Third**, simplify if possible.

EXAMPLES

Divide each of the following and simplify.

1) $\dfrac{\sqrt{72}}{\sqrt{2}}$

$= \sqrt{\dfrac{72}{2}}$

$= \sqrt{36}$

$= 6$

2) $\dfrac{8\sqrt{12}}{2\sqrt{3}}$

$= \dfrac{8}{2}\sqrt{\dfrac{12}{3}}$

$= 4\sqrt{4}$

$= 4 \cdot 2$

$= 8$

3) $\dfrac{8\sqrt{40}}{4\sqrt{5}}$

$= \dfrac{8}{4}\sqrt{\dfrac{40}{5}}$

$= 2\sqrt{8}$

$= 2\sqrt{4 \cdot 2}$

$= 2 \cdot \sqrt{4} \cdot \sqrt{2}$

$= 2 \cdot 2 \cdot \sqrt{2}$

$= 4\sqrt{2}$

4) $\dfrac{\sqrt{21} + \sqrt{35}}{\sqrt{7}}$

$= \dfrac{\sqrt{21}}{\sqrt{7}} + \dfrac{\sqrt{35}}{\sqrt{7}}$

$= \sqrt{\dfrac{21}{7}} + \sqrt{\dfrac{35}{7}}$

$= \sqrt{3} + \sqrt{5}$

5) $\dfrac{\sqrt{6x^3}}{\sqrt{2x}}$

$= \sqrt{\dfrac{6x^3}{2x}}$

$= \sqrt{3x^2}$

$= \sqrt{x^2} \cdot \sqrt{3}$

$= x\sqrt{3}$

6) $\dfrac{15\sqrt{x^5}}{3\sqrt{x^2}}$

$= \dfrac{15}{3}\sqrt{\dfrac{x^5}{x^2}}$

$= 5\sqrt{x^3}$

$= 5\sqrt{x^2 \cdot x}$

$= 5 \cdot \sqrt{x^2} \cdot \sqrt{x}$

$= 5x\sqrt{x}$

8-5 Dividing Radical Expressions

Divide and simplify each of the following.

1) $\dfrac{\sqrt{75}}{\sqrt{3}}$

2) $\dfrac{\sqrt{64}}{\sqrt{8}}$

3) $\dfrac{6\sqrt{9}}{2\sqrt{3}}$

4) $\dfrac{12\sqrt{20}}{3\sqrt{5}}$

5) $\dfrac{20\sqrt{48}}{5\sqrt{6}}$

6) $\dfrac{8\sqrt{48}}{2\sqrt{3}}$

7) $\dfrac{\sqrt{x^3}}{\sqrt{x}}$

8) $\dfrac{6\sqrt{27a^5}}{2\sqrt{3a^3}}$

9) $\dfrac{\sqrt{27}+\sqrt{75}}{\sqrt{3}}$

10) $\dfrac{\sqrt{8}+\sqrt{16}}{\sqrt{4}}$

11) $\dfrac{10\sqrt{48a^3b}}{5\sqrt{3ab}}$

12) $\dfrac{3\sqrt{54}}{6\sqrt{3}}$

13) $\dfrac{20\sqrt{50}}{4\sqrt{2}}$

14) $\dfrac{\sqrt{24}-\sqrt{6}}{\sqrt{2}}$

15) $\dfrac{6\sqrt{27x^5y^2}}{2\sqrt{3x^3}}$

16) $\dfrac{5\sqrt{48x^3y}}{10\sqrt{3xy}}$

17) $\dfrac{\sqrt{27x^3}}{\sqrt{9x}}$

18) $\dfrac{9\sqrt{5x}}{3\sqrt{x}}$

REVIEW

Simplify each of the following.

1) $\sqrt{2500}$

2) $\sqrt{125}$

3) $\sqrt{12x^2}$

4) $\sqrt{75x^4}$

8-6 Rationalizing the Denominator

A fraction that has a **radical** in its **denominator** is not in its simplest form. The process of eliminating a radical from the denominator of a fraction is called **rationalizing the denominator**. To rationalize the denominator, simply multiply both the numerator and the denominator by a number that will make the **denominator** a **whole number**.

Remember the following when simplifying radical expressions.

Helpful Hints
- **First**, rationalize the denominator.
- **Second**, completely simplify the result.
- There is no exact order to be used. Just remember to apply the rules of radicals properly.

EXAMPLES

Rationalize the denominator and simplify completely each of the following.

1) $\dfrac{3}{\sqrt{5}}$

$= \dfrac{3}{\sqrt{5}} \cdot \dfrac{\sqrt{5}}{\sqrt{5}}$

$= \dfrac{3\sqrt{5}}{5}$

2) $\sqrt{\dfrac{1}{3}}$

$= \dfrac{\sqrt{1}}{\sqrt{3}}$

$= \dfrac{\sqrt{1} \cdot \sqrt{3}}{\sqrt{3} \cdot \sqrt{3}}$

$= \dfrac{\sqrt{1 \cdot 3}}{3}$

$= \dfrac{\sqrt{3}}{3}$

3) $\dfrac{\sqrt{5}}{\sqrt{8}}$

$= \dfrac{\sqrt{5}}{\sqrt{8}} \cdot \dfrac{\sqrt{8}}{\sqrt{8}}$

$= \dfrac{\sqrt{40}}{8}$

$= \dfrac{\sqrt{4 \cdot 10}}{8}$

$= \dfrac{\sqrt{4} \cdot \sqrt{10}}{8}$

$= \dfrac{2 \cdot \sqrt{10}}{8}$

$= \dfrac{1}{4} \cdot \sqrt{10}$

$= \dfrac{\sqrt{10}}{4}$

4) $\dfrac{2\sqrt{8}}{\sqrt{2}}$

$= \dfrac{2\sqrt{8}}{\sqrt{2}} \cdot \dfrac{\sqrt{2}}{\sqrt{2}}$

$= \dfrac{2\sqrt{8 \cdot 2}}{2}$

$= \dfrac{2\sqrt{16}}{2}$

$= \dfrac{2}{2} \cdot \sqrt{16}$

$= 1 \cdot \sqrt{16}$

$= 4$

8-6 Rationalizing the Denominator

Rationalize the denominator and simplify completely each of the following.

1) $\sqrt{\dfrac{3}{5}}$

2) $\sqrt{\dfrac{1}{2}}$

3) $\dfrac{\sqrt{7}}{\sqrt{8}}$

4) $\dfrac{12}{\sqrt{2}}$

5) $\dfrac{9}{\sqrt{3}}$

6) $\dfrac{25}{\sqrt{5}}$

7) $\dfrac{3\sqrt{18}}{\sqrt{2}}$

8) $\sqrt{\dfrac{x^2}{2}}$

9) $\dfrac{18}{3\sqrt{2}}$

10) $\dfrac{7}{\sqrt{7}}$

11) $\dfrac{6}{\sqrt{12}}$

12) $\dfrac{\sqrt{5}}{\sqrt{3}}$

13) $\dfrac{36}{\sqrt{18}}$

14) $\dfrac{12}{2\sqrt{3}}$

15) $\dfrac{xy}{\sqrt{y}}$

16) $\dfrac{\sqrt{8}+\sqrt{18}}{\sqrt{2}}$

17) $\dfrac{4}{3\sqrt{8}}$

18) $\dfrac{12}{\sqrt{32}}$

REVIEW

Simplify each of the following.

1) $n^2 = 121$

2) $\dfrac{x}{4} = \dfrac{25}{x}$

3) $x^2 - 5 = 11$

4) $x^2 + 2 = 27$

8-7 Simplifying Radical Expressions with Binomial Denominators

Sometimes it is necessary to **rationalize** a **binomial denominator containing a radical**. These may appear to be complicated, but they are not. Simply multiply the numerator and the denominator by the given denominator with its middle sign changed. This will make the denominator a difference of two squares, and will clear out the radical.

Remember to keep the following in mind when simplifying radical expressions with binomial denominators.

Helpful Hints
- Remove the radical in the denominator by multiplying the numerator and denominator by the same denominator with its middle sign changed. The denominator will then be the difference of the two squares.

- Remember to use the distributive property when necessary.

- Be sure that your final answer is completely simplified.

- Remember that $(a + b)(a - b) = a^2 - b^2$.

- Leave no radicals in the denominator.

- Leave no fractions in the denominator.

EXAMPLES

Rationalize the denominator and simplify each of the following.

1) $\dfrac{2}{\sqrt{2} + 3}$

$= \dfrac{2}{(\sqrt{2} + 3)} \cdot \dfrac{(\sqrt{2} - 3)}{(\sqrt{2} - 3)}$

$= \dfrac{2(\sqrt{2} - 3)}{\sqrt{2}^2 - 3^2}$

$= \dfrac{2\sqrt{2} - 6}{2 - 9}$

$= \dfrac{2\sqrt{2} - 6}{-7}$

2) $\dfrac{\sqrt{10}}{3 + \sqrt{6}}$

$= \dfrac{\sqrt{10}}{3 + \sqrt{6}} \cdot \dfrac{3 - \sqrt{6}}{3 - \sqrt{6}}$

$= \dfrac{\sqrt{10}(3 - \sqrt{6})}{3^2 - \sqrt{6}^2}$

$= \dfrac{3\sqrt{10} - \sqrt{10} \cdot \sqrt{6}}{9 - 6}$

$= \dfrac{3\sqrt{10} - \sqrt{60}}{3}$

$= \dfrac{3\sqrt{10} - \sqrt{4 \cdot 15}}{3}$

$= \dfrac{3\sqrt{10} - \sqrt{4} \cdot \sqrt{15}}{3}$

$= \dfrac{3\sqrt{10} - 2\sqrt{15}}{3}$

8-7 Simplifying Radical Expressions with Binomial Denominators

EXERCISES

Rationalize the denominator and simplify each of the following. Remember to express your answers in their simplest form. You may need to cancel out common factors.

1) $\dfrac{1}{\sqrt{3}+1}$

2) $\dfrac{5}{2+\sqrt{3}}$

3) $\dfrac{\sqrt{6}}{1-\sqrt{6}}$

4) $\dfrac{\sqrt{5}}{2+\sqrt{6}}$

5) $\dfrac{20}{\sqrt{6}+2}$

6) $\dfrac{24}{\sqrt{15}-3}$

7) $\dfrac{26}{4-\sqrt{3}}$

8) $\dfrac{\sqrt{2}}{4-\sqrt{3}}$

9) $\dfrac{5}{\sqrt{2}+3}$

10) $\dfrac{14}{3-\sqrt{2}}$

11) $\dfrac{2\sqrt{7}}{\sqrt{7}-1}$

12) $\dfrac{6\sqrt{2}}{2-\sqrt{5}}$

13) $\dfrac{\sqrt{3}-1}{\sqrt{3}+1}$

14) $\dfrac{2-\sqrt{7}}{\sqrt{7}-4}$

15) $\dfrac{1}{\sqrt{5}+1}$

REVIEW

Simplify each of the following.

1) $4\sqrt{2}+5\sqrt{2}$

2) $\sqrt{27}+\sqrt{3}$

3) $\sqrt{27a}-2\sqrt{3a}$

4) $\sqrt{12}+\sqrt{3}$

8-8 The Pythagorean Theorem

The **Pythagorean theorem** is an important formula. It was named after Pythagoras, a Greek who lived in the Sixth Century.

The theorem states that in any **right triangle**, the square of the length of the **hypotenuse** is equal to the sum of the squares of the lengths of the other two sides.

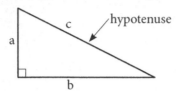

$$a^2 + b^2 = c^2$$

When you know the lengths of two sides of a right triangle, you can find the length of the third side.

You need to know this theorem since it is one of the most useful laws in mathematics. It involves the use of exponents and square roots. Remember the following when using the Pythagorean theorem.

Helpful Hints
- **First**, draw a sketch of the triangle.
- **Second**, write out the formula
- **Third**, substitute the values into the formula.
- **Fourth**, solve for the missing side.
- If you want your answer in the form of a decimal, use a calculator or a square root table.

EXAMPLES

Find the length of the missing side.

1)

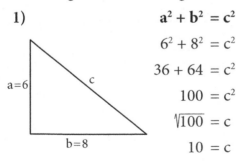

$$a^2 + b^2 = c^2$$
$$6^2 + 8^2 = c^2$$
$$36 + 64 = c^2$$
$$100 = c^2$$
$$\sqrt{100} = c$$
$$10 = c$$

2)

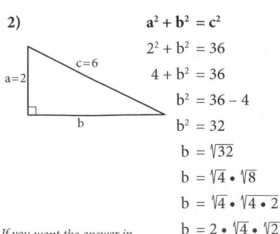

$$a^2 + b^2 = c^2$$
$$2^2 + b^2 = 36$$
$$4 + b^2 = 36$$
$$b^2 = 36 - 4$$
$$b^2 = 32$$
$$b = \sqrt{32}$$
$$b = \sqrt{4} \cdot \sqrt{8}$$
$$b = \sqrt{4} \cdot \sqrt{4 \cdot 2}$$
$$b = 2 \cdot \sqrt{4} \cdot \sqrt{2}$$
$$b = 2 \cdot 2 \cdot \sqrt{2} = 4\sqrt{2}$$

If you want the answer in decimal form, either use a square root table or a calculator.

Chapter 8: **RADICAL EXPRESSIONS AND GEOMETRY**

8-8 The Pythagorean Theorem

Find the length of the missing side and simplify your answer.

1)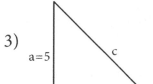
   a=2, b=4, c

2)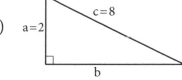
   c=14, a=12, b

3)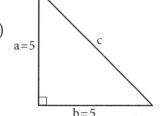
   a=5, b=5, c

4)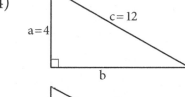
   c=12, a=4, b

5)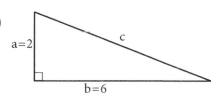
   a=2, b=6, c

6)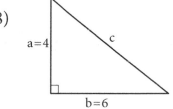
   c=8, a=2, b

7)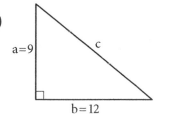
   a=9, b=12, c

8)
   a=4, b=6, c

Simplify each of the following.

1) $\dfrac{\sqrt{21}}{\sqrt{7}}$

2) $\dfrac{\sqrt{a^3}}{\sqrt{a}}$

3) $\dfrac{\sqrt{8x}}{\sqrt{4x}}$

4) $\sqrt{\dfrac{10}{4}} \cdot \sqrt{\dfrac{4}{2}}$

8-9 The Distance Formula

INTRODUCTION

The **Distance Formula** allows you to find the distance between any two points on the coordinate plane. When using this formula, you are actually using the **Pythagorean theorem**. Here is the Distance Formula.

$$D = \sqrt{(x_2 - x_1)^2 + (y_2 - y_1)^2}$$

When using the Distance Formula remember the following.

Helpful Hints

- **First**, draw a sketch and plot the points on the coordinate plane.
- **Second**, label each point as Point$_1$ and Point$_2$.
- **Third**, write out the formula and substitute the coordinate values of each point into the formula and solve.
- **Fourth**, simplify your answer. Use a calculator to get the answer as a decimal if necessary.

EXAMPLES

Use the Distance Formula to find the distance between the given points.

Find the distance between P$_2$ (5, 4) and P$_1$ (-2, 1).

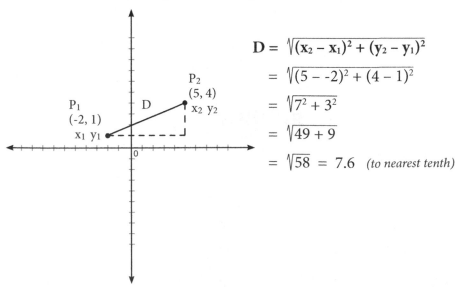

$$D = \sqrt{(x_2 - x_1)^2 + (y_2 - y_1)^2}$$
$$= \sqrt{(5 - -2)^2 + (4 - 1)^2}$$
$$= \sqrt{7^2 + 3^2}$$
$$= \sqrt{49 + 9}$$
$$= \sqrt{58} = 7.6 \quad \textit{(to nearest tenth)}$$

If you look carefully at the sketch, you can see that the distance = D. It represents the hypotenuse of the right triangle which was formed.

8-9 The Distance Formula

Draw a sketch and use the Distance Formula to find the distance between the given points. Round each answer to the nearest tenth. $D = \sqrt{(x_2 - x_1)^2 + (y_2 - y_1)^2}$

1) (4, 3) (2, 1)

2) (2, 2) (3, 3)

3) (1, 1) (4, 2)

4) (-7, -9) (2, -9)

5) (-5, -3) (-9, -6)

6) (10, 3) (-4, 9)

7) (-4, 6) (5, 2)

8) (-4, 7) (-9, -5)

9) (2, -3) (8, 7)

10) (-2, 1) (-5, -4)

11) (10, 5) (-7, 0)

12) (8, -2) (4, -6)

13) (-1, -1) (-9, -7)

14) (-4, -2) (-8, -5)

REVIEW

Simplify by rationalizing the denominator.

1) $\sqrt{\dfrac{3}{4}}$

2) $\sqrt{\dfrac{5}{8}}$

3) $\dfrac{3\sqrt{18}}{\sqrt{2}}$

4) $\dfrac{6}{\sqrt{12}}$

INTRODUCTION

The **Midpoint Formula** allows you to find the **coordinates of the point** that is located halfway between any two points on the coordinate plane. The name for this point is the **midpoint**. Here is the Midpoint Formula.

$$M = \left(\frac{x_1 + x_2}{2}\right), \left(\frac{y_1 + y_2}{2}\right)$$

When using the Midpoint Formula remember the following.

Helpful Hints
- **First**, draw a sketch and plot the points on the coordinate plane.
- **Second**, label each point as $Point_1$ and $Point_2$.
- **Third**, write out the formula and substitute the coordinate values of each point into the formula and solve.

EXAMPLE

Find the coordinates of the midpoint between the two given points.

(5, 2) (2, -3)

$$M = \left(\frac{x_1 + x_2}{2}\right), \left(\frac{y_1 + y_2}{2}\right)$$

$$= \left(\frac{5 + 2}{2}\right), \left(\frac{2 + {-3}}{2}\right)$$

$$= \left(\frac{7}{2}\right), \left(\frac{-1}{2}\right)$$

The coordinates of the midpoint are

$$\left(\frac{7}{2}, \frac{-1}{2}\right)$$

(also can be written as)

$$\left(3\tfrac{1}{2}, -\tfrac{1}{2}\right)$$

Remember $\frac{-1}{2} = \frac{1}{-2} = -\frac{1}{2}$

On the graph:
P₂ (5, 2) $x_2\ y_2$
$\left(3\tfrac{1}{2}, -\tfrac{1}{2}\right)$
P₁ (2, -3) $x_1\ y_1$

8-10 The Midpoint Formula

Find the coordinates of the midpoint between the two given points. $M = (\frac{x_1 + x_2}{2}), (\frac{y_1 + y_2}{2})$

1) $(4, 2)$ $(8, 5)$ 2) $(2, -3)$ $(7, 6)$

3) $(-3, 4)$ $(5, -3)$ 4) $(6, 1)$ $(-6, -2)$

5) $(2, 0)$ $(5, 0)$ 6) $(7, -3)$ $(-3, -2)$

7) $(5, 8)$ $(7, -3)$ 8) $(-5, -5)$ $(6, 4)$

9) $(-3, 2)$ $(5, 4)$ 10) $(-3, -5)$ $(7, -2)$

11) $(3, 7)$ $(5, 5)$ 12) $(8, 5)$ $(-5, 6)$

13) $(4, -12)$ $(-8, 1)$ 14) $(-7, 10)$ $(-7, 6)$

REVIEW

Use the Pythagorean Theorem $a^2 + b^2 = c^2$ to find the missing value. Remember to draw a sketch.

1) $a = 3, b = 4, c = ?$ 2) $a = 9, b = 12, c = ?$

3) $a = 6, b = ?, c + 10$ 4) $a = ?, b = 20, c = 25$

Chapter 8 Review: Radical Expressions and Geometry

Simplify.

1) $\sqrt{900}$

2) $\sqrt{5x^2y^2}$

3) $\sqrt{100x^3}$

4) $\sqrt{x^2 - 10x + 25}$

Solve.

5) $x^2 = 121$

6) $x^2 - 5 = 11$

Add, subtract, multiply, or divide each of the following.

7) $5\sqrt{12} - 2\sqrt{3}$

8) $3\sqrt{27} + 3\sqrt{12}$

9) $5\sqrt{3} - 4\sqrt{27}$

10) $2\sqrt{2} \cdot \sqrt{3} \cdot \sqrt{12}$

11) $(6\sqrt{5})^2$

12) $\dfrac{5\sqrt{16x^3}}{8\sqrt{25y^2}}$

Simplify by rationalizing the denominator.

13) $\sqrt{\dfrac{3}{8}}$

14) $\dfrac{2\sqrt{8}}{\sqrt{2}}$

Chapter 8 Review: Radical Expressions and Geometry

Solve each equation.

15) $\sqrt{3x} = 6$

16) $\sqrt{4x - 3} = 5$

17) A right triangle has one leg of 9 units and the other leg of 12 units. Find the length of the hypotenuse.

For questions 18–19, use the distance formula to find the distance between each pair of points.

$$D = \sqrt{(x_2 - x_1)^2 + (y_2 - y_1)^2}$$

18) $(2, -9), (-7, -9)$

19) $(4, 3), (2, 7)$

For questions 20–21, use the mid-point formula to find the mid-point for each pair of points.

$$\textbf{Mid-Point} = \left(\frac{x_1 + x_2}{2}\right), \left(\frac{y_1 + y_2}{2}\right)$$

20) $(4, -12), (-8, 1)$

21) $(-4, -2), (-8, -5)$

9-1 Factoring Quadratic Equations

INTRODUCTION

A **quadratic equation** is an equation in one unknown that has the highest degree 2. The **standard form** of a quadratic equation is $ax^2 + bx + c = 0$

The following are examples of quadratic equations:

$$x^2 - 5x + 4 = 0 \qquad 3x^2 + 2x = 0 \qquad x^2 = 5$$

You will be working with **complete** as well as **incomplete** quadratic equations. A **complete quadratic equation** contains the first and second-degree terms of the unknown as well as a constant. Example: $x^2 + 7x + 6 = 0$

An **incomplete quadratic equation** is missing either the term with the first power of the unknown **or** the constant. Examples: $x^2 - 3x = 0 \qquad x^2 = 8$

Factoring is useful for solving many quadratic equations. Remember the following when solving quadratic equations using factoring.

Helpful Hints
- **First**, if necessary put the equation in standard form. This may involve removing parentheses, clearing out fractions, and combining like terms to the left of the equal sign. Make the right side of the equal sign zero.
- **Second**, factor the left side of the equal sign.
- **Third**, write two equations with each factor equal to zero.
- **Fourth**, solve each equation. Check by substituting each value into the original equation.
- It might be good to review the lesson on factoring polynomials.

EXAMPLES

Solve each of the following.

1) $x^2 - 4x = 5$

$x^2 - 4x - 5 = 0$ *Add -5 to both sides to put in standard form*

$(x - 5)(x + 1) = 0$ *Factor*

$x - 5 = 0 \quad$ or $\quad x + 1 = 0$ *Set each factor = 0*

$x = 5$ or $x = -1$ *Solve*

2) $x^2 = -4x$

$x^2 + 4x = 0$ *Rewrite*

$x(x + 4) = 0$ *Factor*

$x = 0 \quad$ or $\quad x + 4 = 0$ *Set each factor = 0*

$x = 0$ or $x = -4$

9-1 Factoring Quadratic Equations

3) $x(x + 3) = 40$

   $x^2 + 3x = 40$ *Multiply the left side*

   $x^2 + 3x - 40 = 0$ *Put in standard form*

   $(x + 8)(x - 5) = 0$ *Factor*

   $x + 8 = 0$ or $x - 5 = 0$

   $x = -8$ or $x = 5$

4) $\dfrac{x}{4} = \dfrac{7}{x + 3}$

   $x(x + 3) = 4 \bullet 7$ *Cross multiply*

   $x^2 + 3x = 28$

   $x^2 + 3x - 28 = 0$ *Put in standard form*

   $(x + 7)(x - 4) = 0$ *Factor*

   $x + 7 = 0$ or $x - 4 = 0$

   $x = -7$ or $x = 4$

EXERCISES

Solve each of the following.

1) $x^2 - 6x + 8 = 0$

2) $x^2 + 5x = 6$

3) $x^2 - 13x + 22 = 0$

4) $x^2 + x = 6$

5) $x^2 = 5x$

6) $x^2 = -4x$

7) $x^2 - 49 = 0$

8) $x^2 = 5x + 24$

9) $\dfrac{x}{2} = \dfrac{6}{x + 1}$

10) $x(x + 4) = 12$

11) $\dfrac{4}{x} = \dfrac{x - 1}{3}$

12) $x^2 - 8x + 12 = 0$

13) $x^2 - 36 = 0$

14) $x^2 = 27 - 6x$

15) $x(x - 3) = 28$

16) $x^2 - 9x + 18 = 2x$

17) $x - 2 = \dfrac{3}{x}$

18) $x^2 - 2x - 3 = 0$

REVIEW

Use the Pythagorean Theorem to find the missing part of each right triangle ABC. Remember to draw a sketch.

1) leg a = 4, leg b = 12, hypotenuse = ?

2) leg a = 2, leg b = ?, hypotenuse = 8

9-2 Solving Quadratic Equations Using Square Roots

Sometimes an **incomplete quadratic equation** is missing the **first-degree term**. Sometimes they are not written in standard form. A good way to solve these is by using square roots.

Remember the following when working with this type of quadratic equation.

Helpful Hints
- There will be **two** solutions, **one positive** and **one negative**.
- Sometimes it is necessary to eliminate fractions or combine like terms.
- Use the skills that you have learned to solve the equations. If you feel comfortable, do some of the steps mentally.

EXAMPLES

Solve each of the following.

1) $x^2 - 25 = 0$

$x^2 = 25$ *Rewrite*

$\sqrt{x^2} = \sqrt{25}$ *Take the square root of each side*

$x = \pm 5$

$x = 5, -5$

2) $3x^2 = 27$

$\dfrac{3x^2}{3} = \dfrac{27}{3}$ *Divide each side by 3*

$x^2 = 9$

$\sqrt{x^2} = \sqrt{9}$

$x = \pm 3$

$x = 3, -3$

3) $4x^2 - 13 = x^2 + 14$

$4x^2 - x^2 = 14 + 13$ *Collect like terms and rewrite the equation*

$3x^2 = 27$

$\dfrac{3x^2}{3} = \dfrac{27}{3}$ *Divide each side by 3*

$x^2 = 9$

$\sqrt{x^2} = \sqrt{9}$

$x = \pm 3$

$x = 3, -3$

4) $3(2x + 2)^2 = 48$

$\dfrac{3(2x + 2)^2}{3} = \dfrac{48}{3}$ *Divide each side by 3*

$(2x + 2)^2 = 16$

$\sqrt{(2x + 2)^2} = \sqrt{16}$ *Take the square root of each side*

$2x + 2 = \pm 4$

$2x + 2 = 4$ —or— $2x + 2 = -4$

$2x = 2$ $2x = -6$

$x = 1$ $x = -3$

$x = 1, -3$

9-2 Solving Quadratic Equations Using Square Roots

EXERCISES

Solve each of the following.

1) $x^2 - 81 = 0$

2) $x^2 = 64$

3) $5x^2 = 45$

4) $2x^2 - 8 = 0$

5) $x^2 - 11 = 70$

6) $\frac{x}{9} = \frac{4}{x}$

7) $3x^2 - 60 = 0$

8) $(x - 4)^2 = 25$

9) $(4x - 2)^2 = 36$

10) $2(x - 7)^2 = 18$

11) $2(2y + 4)^2 = 72$

12) $5x^2 = 30$

13) $4x^2 + 5 = 21$

14) $3x^2 + 4x^2 = 35$

15) $(x + 2)^2 = 16$

16) $(x - 3)^2 = 49$

REVIEW

Use the distance formula, $D = \sqrt{(x_2 - x_1)^2 + (y_2 - y_1)^2}$, to find the distance between the given points.

1) $(-1, -1)\ (-9, -7)$

2) $(-4, -2)\ (-8, -5)$

9-3 Completing the Square

INTRODUCTION

Completing the square is another method that can be used to solve a quadratic equation. It can be especially useful when the equation cannot be factored.

It is easy to solve a trinomial quadratic equation that is a perfect square. Even if a quadratic equation is not in this form, sometimes it is possible to transform it into one that is. Remember these steps when completing the square.

Helpful Hints
- **First**, set the equation up in the form $ax^2 + bx = c$, where c is a constant
- **Second**, take half the coefficient of x.
- **Third**, square that answer.
- **Fourth**, add the result to both sides of the equal sign.
- **Fifth**, you now have a perfect square on the left side of the equal sign that you can easily solve.
- Sometimes there will be **radicals** in your answers.

EXAMPLES

Solve each of the following by completing the square.

1) $x^2 - 10x + 21 = 0$

$x^2 - 10x = -21$ *Rewrite*

$\frac{1}{2}$ **of -10 = -5, (-5)^2 = 25**
   Add 25 to both sides

$x^2 - 10x + 25 = -21 + 25$

$(x - 5)^2 = 4$ *Factor*

$\sqrt{(x-5)^2} = \sqrt{4}$ *Take the square root of each side*

$x - 5 = \pm 2$

$x - 5 = 2$ or $x - 5 = -2$

$x = 7$ or $x = 3$

The solutions are 7 and 3

2) $x^2 - 6x + 4 = 0$

$x^2 - 6x = -4$ *Rewrite*

$\frac{1}{2}$ **of -6 = -3, (-3)^2 = 9**
   Add 9 to both sides

$x^2 - 6x + 9 = -4 + 9$

$(x - 3)^2 = 5$ *Factor*

$\sqrt{(x-3)^2} = \sqrt{5}$ *Take the square root of each side*

$x - 3 = \pm\sqrt{5}$

$x - 3 = \sqrt{5}$ or $x - 3 = -\sqrt{5}$

$x = 3 + \sqrt{5}$ or $x = 3 - \sqrt{5}$

The solutions are $3 + \sqrt{5}$ and $3 - \sqrt{5}$

9-3 Completing the Square

EXERCISES

Solve each of the following by completing the square.

1) $x^2 + 6x + 5 = 0$

2) $x^2 - 3x - 18 = 0$

3) $x^2 - 6x - 7 = 0$

4) $x^2 - 4x = 5$

5) $x^2 - 2x = 4$

6) $x^2 - 4x - 12 = 0$

7) $x^2 + 4x = -2$

8) $x^2 + 2x - 6 = 0$

9) $x^2 + 4x + 1 = 0$

10) $x^2 + 6x + 1 = 0$

11) $x^2 + 2x - 1 = 0$

12) $x^2 - 4x - 1 = 0$

13) $x^2 - 6x - 9 = 0$

14) $x^2 - 4x = 1$

REVIEW

Use the mid-point formula, $M = \left(\frac{x_1 + x_2}{2}\right), \left(\frac{y_1 + y_2}{2}\right)$, to find the mid-point of the following.

1) $(3, 2), (7, 8)$

2) $(2, -3) (8, 7)$

3) $(5, 3) (-1, 4)$

4) $(12, 2) (-6, 7)$

9-4 The Quadratic Formula

The **quadratic formula** can be used to solve any quadratic equation. This is one of the most important formulas in all of mathematics, and it should be memorized.

$$x = \frac{-b \pm \sqrt{b^2 - 4ac}}{2a}$$

Remember the following steps when using the quadratic formula to solve an equation.

Helpful Hints
- **First**, write the standard form of the quadratic equation $ax^2 + bx + c = 0$.
- **Second**, write the equation to be solved in standard form, directly below.
- **Third**, identify the values for a, b, and c.
- **Fourth**, substitute these values into the quadratic formula.
- **Fifth**, simplify by completing the necessary steps.
- At first, it is good to have the quadratic formula in front of you until you are sure that you have memorized it.
- Check your answers by substituting them into the original equation.

EXAMPLES

Use the quadratic formula to solve each of the following.

1) $2x^2 + 9x - 5 = 0$

$ax^2 + bx + c = 0$

$2x^2 + 9x - 5 = 0$

$a = 2, b = 9, c = -5$

$x = \frac{-b \pm \sqrt{b^2 - 4ac}}{2a}$

$x = \frac{-9 \pm \sqrt{(9)^2 - 4(2)(-5)}}{2(2)}$

$x = \frac{-9 \pm \sqrt{81 + 40}}{4}$

$x = \frac{-9 \pm \sqrt{121}}{4}$

$x = \frac{-9 + 11}{4}$ or $x = \frac{-9 - 11}{4}$

$x = \frac{2}{4}$ $x = \frac{-20}{4}$

$x = \frac{1}{2}$ or $x = -5$

2) $2x^2 - 12x + 14 = 0$

$ax^2 + bx + c = 0$

$2x^2 - 12x + 14 = 0$

$a = 2, b = -12, c = 14$

$x = \frac{-b \pm \sqrt{b^2 - 4ac}}{2a}$

$x = \frac{12 \pm \sqrt{(-12)^2 - 4(2)(14)}}{2(2)}$

$x = \frac{12 \pm \sqrt{144 - 112}}{4}$

$x = \frac{12 \pm \sqrt{32}}{4}$

$x = \frac{12 \pm 4\sqrt{2}}{4}$ *Simplify*

$x = \frac{\cancel{12}^{3}}{\cancel{4}_{1}} \pm \frac{\cancel{4}^{1}\sqrt{2}}{\cancel{4}_{1}}$ *Can call out common factors*

$x = 3 \pm \sqrt{2}$

$x = 3 + \sqrt{2}$ or $x = 3 - \sqrt{2}$

9-4 The Quadratic Formula

Use the quadratic formula to solve each of the following. $x = \dfrac{-b \pm \sqrt{b^2 - 4ac}}{2a}$

1) $2x^2 - 5x + 2 = 0$

2) $2x^2 + 3x + 1 = 0$

3) $x^2 + 2x - 8 = 0$

4) $x^2 - 3x + 2 = 0$

5) $x^2 - 2x - 3 = 0$

6) $5x^2 + 7x + 2 = 0$

7) $2x^2 + x - 7 = 0$

8) $x^2 - 5x - 14 = 0$

9) $2x^2 + 3x - 1 = 0$

10) $x^2 + 3x + 1 = 0$

11) $2x^2 - 3x - 2 = 0$

12) $x^2 - x - 1 = 0$

13) $2x^2 + 5x + 3 = 0$

14) $x^2 + 4x - 5 = 0$

REVIEW

Solve each of the following by factoring.

1) $y^2 - 3y = 0$

2) $16x^2 - 4x^2 = 0$

3) $x^2 + 12x + 36 = 0$

4) $36x^2 - 36 = 0$

9-5 Graphing Quadratic Equations

INTRODUCTION

The graph of a quadratic equation is a **parabola**. The tip of the parabola is called the **vertex**. If the parabola opens **upward**, the vertex is called the **minimum**. If it opens **downward** the vertex is called the **maximum**. The **axis of symmetry** goes through the center and divides the parabola into **two mirror images**.

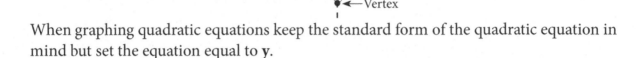

When graphing quadratic equations keep the standard form of the quadratic equation in mind but set the equation equal to **y**.

$$y = ax^2 + bx + c$$

The following is a **4-step process** that can be easily used to graph a quadratic equation.

Helpful Hints

- **First**, determine the **maximum** or **minimum**. If **a** is positive, the graph opens upward and will have a minimum. If **a** is negative, the graph will open downward and will have a maximum.

- **Second**, determine the **axis of symmetry**. The axis of symmetry is found by using the formula $x = \dfrac{-b}{2a}$.

- **Third**, find the **coordinates of the vertex**. This is done by substituting the value of **x** from the second step into the original equation.

- **Fourth**, choose some other values for **x** and solve for the corresponding **y** values. Then plot the order pairs and sketch the parabola.

EXAMPLE

Use the 4-step process to graph the following equation: $y = 2x^2 + 4x + 1$

1) Since **a = +2**, the parabola will open **upward**.

2) The axis of symetry is $x = \dfrac{-4}{2(2)} = -1$

3) To get the coordinate of the vertex, substitute x = -1 into the original equation and solve for y.

$y = 2x^2 + 4x + 1$
$y = 2(-1)^2 + 4(-1) + 1$
$y = 2 + -4 + 1$
y = -1

The coordinates of the vertex are (-1, -1)

4) Select some values for x, and solve for y by substituting into the original equation. Then plot the points.

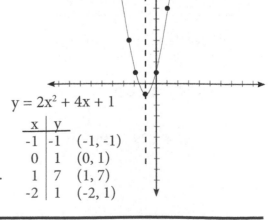

$y = 2x^2 + 4x + 1$

x	y	
-1	-1	(-1, -1)
0	1	(0, 1)
1	7	(1, 7)
-2	1	(-2, 1)

9-5 Graphing Quadratic Equations

Use the 4-step process to graph each of the following.

1) $y = x^2 - 6x + 8$

2) $y = -x^2 + 2x - 3$

3) $y = 2x^2 - 4x - 1$

4) $y = 2x^2 - 8x + 3$

5) $y = x^2 + 2x$

6) $y = -2x^2 + 4x + 1$

7) $y = -3x^2 + 6x + 2$

8) $y = 4x^2$

9) $y = 4x^2 + 8x - 3$

10) $y = x^2 + 2x + 1$

REVIEW

Solve each of the following using square roots.

1) $2x^2 = 50$

2) $x^2 = 8$

3) $y^2 - 15 = 12$

4) $2n^2 - 7 = 13$

9-6 The Discriminant

INTRODUCTION

The quadratic formula is

$$x = \frac{-b \pm \sqrt{b^2 - 4ac}}{2a}$$

The expression inside the square root sign, $b^2 - 4ac$, is called the **discriminant**. The discrminant can tell you how many times the graph of a quadratic function crosses the x-axis. The graph will cross **once**, **twice**, or **not at all**. The discriminant can be a useful tool when graphing quadratic functions. Remember the following when using the discriminant.

Helpful Hints
- Keep the standard form $y = ax^2 + bx + c$ in mind.
- If $b^2 - 4ac < 0$, the graph crosses the x-axis 0 times.
- If $b^2 - 4ac > 0$, the graph crosses the x-axis 2 times.
- If $b^2 - 4ac = 0$, the graph crosses the x-axis 1 time.

EXAMPLES

Determine how many times the graph of each quadratic function passes through the x-axis by using the discriminant $b^2 - 4ac$.

1) $y = 2x^2 - 4x + 2$

$a = 2, b = -4, c = 2$

$b^2 - 4ac = (-4)^2 - 4(2)(2)$

$= 16 - 16$

$= 0$

The graph crosses the x-axis 1 time.

2) $y = x^2 - 5x - 6$

$a = 1, b = -5, c = -6$

$b^2 - 4ac = (-5)^2 - 4(1)(-6)$

$= 25 - -24$

$= 25 + 24$

$= 49$

The graph crosses the x-axis 2 times.

3) $y = 2x^2 + 4$

$a = 2, b = 0, c = 4$

$b^2 - 4ac = (0)^2 - 4(2)(4)$

$= 0 - 32$

$= -32$

The graph crosses the x-axis 0 times.

4) $y = -2x^2$

$a = -2, b = 0, c = 0$

$b^2 - 4ac = (0)^2 - 4(-2)(0)$

$= 0 - 0$

$= 0$

The graph crosses the x-axis 1 time.

9-6 The Discriminant

Determine how many times the graph of each quadratic function passes through the x-axis by using the discriminant $b^2 - 4ac$.

1) $y = x^2 - 4x - 3$

2) $y = x^2 + 2x + 1$

3) $y = x^2 - 3x + 6$

4) $y = x^2 - 5x + 2$

5) $y = x^2 + x - 6$

6) $y = 3x^2 - 12$

7) $y = -2x^2 + 4x + 1$

8) $y = 3x^2 - 5x - 1$

9) $y = 2x^2 - 4$

10) $y = x^2 + 4$

11) $y = x^2 + 2x + 2$

12) $y = x^2 - 2x$

13) $y = x^2 - 1$

14) $y = -x^2$

15) $y = x^2 - 4x + 2$

16) $y = -3x^2$

REVIEW

Factor each of the following.

1) $36x^2 - 81y^2$

2) $10x^2 + 25x$

3) $x^2 + 7x + 12$

4) $72x^2 - 32$

Chapter 9 Review: Quadratic Equations

Solve by factoring.

    1) $2x^2 + 6x = 0$ 2) $3x^2 + 18x = 0$

    3) $49x^2 - 36 = 0$ 4) $x^2 - 3x - 10 = 0$

    5) $x^2 - 8x + 7 = 0$ 6) $x^2 + 11x - 80 = 0$

Solve by taking the square root.

    7) $y^2 + 4 = 20$ 8) $3x^2 = 48$

    9) $3y^2 + 5 = 50$ 10) $\frac{y}{4} = \frac{16}{y}$

Solve by completing the square.

    11) $m^2 + 4m = 12$ 12) $x^2 - 2x - 4 = 0$

    13) $x^2 + 2x - 6 = 0$

Solve using the Quadratic Formula. $x = \dfrac{-b \pm \sqrt{b^2 - 4ac}}{2a}$

    14) $2x^2 + 3x - 9 = 0$ 15) $x^2 - 6x + 2 = 0$

Use the discriminant to find the number of roots. $D = b^2 - 4ac$

    16) $x^2 - 3x + 6$ 17) $-2x^2 + 4x + 1$

Draw a graph using the 4-step process.

    18) $y = x^2 + 2x + 1$

Keep These Tips in Mind When Solving Algebra Word Problems

Many students find that algebra word problems can be quite challenging. However, you will find that they are really quite easy. The secret is to solve them using an organized plan. There are several types of algebra word problems, and here are a few general tips to keep in mind when solving them.

- First, read the problem carefully and be sure that you fully understand it. Be sure you understand that which is given, and what is to be found.

- Second, select a variable to represent one of the unknowns. The variable will be used to describe all the other unknowns in the problem. Often it is good to have the variable represent the smallest number in the problem.

- Third, translate the word problem into an equation.

- Fourth, solve the equation and use the solution to answer the question that was asked for in the problem. Sometimes the answer will be the value of the variable. Sometimes it will be necessary to use the value of the variable to find what was asked for.

- Fifth, check your answers.

- You will find that sometimes using charts can be quite helpful when solving algebra word problems.

10-1 Introduction to Algebra Word Problems

INTRODUCTION

To solve **algebra word problems**, it is necessary to translate words into **algebraic expressions** containing a **variable**. Keep in mind that a variable is a letter that represents a number. In algebra word problems, the object is to find the **value** of the variable. To do this, you will put the algebraic skills that you have learned to use. Remember the following when translating words into algebraic expressions.

Helpful Hints
- Some of the most common words indicating **addition**: sum, add, more, greater, increased, more than.

- Some of the most common words indicating **subtraction**: minus, less, less than, difference, decreased, diminished, reduced.

- Some of the most common words indicating **multiplication**: product, times, multiplied by.

- Some of the most common words indicating **division**: divide, quotient, ratio.

- "**Is**" often means "equal" (=).

EXAMPLES

To solve **algebra word problems**, it is necessary to translate words into **algebraic expressions** containing a **variable**. A **variable** is a letter that represents a number. Here are some examples:

Three more than a number $\rightarrow$ **x + 3**   Four less than a number $\rightarrow$ **x − 4**

Twice a number $\rightarrow$ **2x**   Seven times a number $\rightarrow$ **7x**

The quotient of x and five $\rightarrow \frac{x}{5}$ A number decreased by six $\rightarrow$ **x − 6**

Seven less than three times a number $\rightarrow$ **3x − 7**   One-third a number $\rightarrow \frac{1}{3}x$ or $\frac{x}{3}$

Twice a number less nine is equal to 15 $\rightarrow$ **2x − 9 = 15**   Two-fifths a number $\rightarrow \frac{2}{5}x$ or $\frac{2x}{5}$

The difference between three times a number and eight equals 12 $\rightarrow$ **3x − 8 = 12**

The sum of a number and -9 is 24 $\rightarrow$ **x + -9 = 24**

Three times a number less six equals twice the number plus 15 $\rightarrow$ **3x − 6 = 2x + 15**

Twice the sum of n and five $\rightarrow$ **2(n + 5)**

The difference between four times x and 15 equals twice the number $\rightarrow$ **4x - 15 = 2x**

10-1 Introduction to Algebra Word Problems

Translate each of the following into an equation.

1) Seven less than twice a number is 12.

2) Two more than three times a number equals 30.

3) The sum of twice a number and five is 14.

4) The difference between four times a number and six is 10.

5) Twelve is five less than four times a number.

6) One-third times a number less four equals twice the number added to eight.

7) Twice the sum of a number and two equals 10.

8) The difference between five times a number and three is 17.

9) Twice a number decreased by six is 15.

10) Two less than three times a number is seven more than twice the number.

11) Four more than a number equals the sum of seven and -12.

12) A number divided by five is 25.

13) Twenty-five is nine more than four times a number.

14) Sixteen subtracted from five times a number is equal to five plus the number.

15) The sum of four x and three is the same as the difference of two x and two.

16) Twice the quantity of two y and 7 equals 10.

17) Four times the quantity of x plus 9 is the same as the difference of x and eight.

18) Fourteen equals the sum of six and a number divided by three.

19) If the product of five and a number is divided by 8, the result is 12.

20) Four times a number less 7 is equal to 12 more than two times that number.

REVIEW

Solve each of the following by factoring.

1) $x^2 - 8x + 15 = 0$ 2) $x^2 - 81 = 0$

3) $2x^2 - 10x + 8 = 0$ 4) $x^2 - 5x = 0$

10-2 Everyday Algebra Word Problems

INTRODUCTION

To solve an everyday **algebra word problem**, the basic idea is to **translate** the words into an **equation**. Once you have the equation, all you need to do is solve it and then check your answer. Use the following steps when solving everyday algebra word problems.

Helpful Hints
- **First**, read the problem **carefully** and be sure that you fully understand it. Be sure you understand that which is **given**, and what is to be **found**.

- **Second**, select a variable to represent one of the unknowns. This variable will be used to describe all the other numbers in the problem. Often it is good to have the variable represent the smallest number in the problem.

- **Third**, translate the problem into an equation.

- **Fourth**, solve the equation and use the solution to answer the question that was asked in the problem. Sometimes the answer will be the value of the variable. Sometimes it will be necessary to use the value of the variable to find what was asked for in the problem.

- **Fifth**, check your answer.

EXAMPLES

Solve each of the algebra word problems using a variable and an equation.

1) The difference between three times a number and 9 is 15. Find the number.

Select the variable	let x = the number
Write the equation	$3x - 9 = 15$
Solve the equation	$3x - 9 + 9 = 15 + 9$
	$3x = 24$
	$x = 8$

The number is 8. *Check your answer.*

2) Four times a number less six is eight more than two time the number. Find the number.

Select the variable	let x = the number
Write the equation	$4x - 6 = 2x + 8$
Solve the equation	$4x - 2x - 6 = 2x - 2x + 8$
	$2x - 6 + 6 = 8 + 6$
	$2x = 14$
	$x = 7$

The number is 7. *Check your answer.*

3) A board 44 cm long is cut into two pieces. The long piece is three times the length of the short piece. What is the length of each piece?

Select the variable	let x = the short piece
	3x = the long piece
Write the equation	$3x + x = 44$
Solve the equation	$4x = 44$
	$x = 11$

The short piece, x, is 11 cm.
The long piece, 3x, is 33 cm. *Check your answers.*

4) Roy weighs 50 kg more than Bill. Their combined weight is 170 kg. What is each of their weights?

Select the variable	let x = Bill's weight
	x + 50 = Roy's weight
Write the equation	$x + (x + 50) = 170$
Solve the equation	$2x + 50 - 50 = 170 - 50$
	$2x = 120$
	$x = 60$

Bill's weight, x, is 60 kg.
Roy's weight, x + 50, is 110 kg. *Check your answers.*

5) Find two consecutive integers whose sum is 91.

Select the variables	let x = first integer
	(x + 1) = second integer
Write the equation	x + (x + 1) = 91
Solve the equation	2x + 1 = 91
	2x + 1 −1 = 91 − 1
	2x = 90
	x = 45

The first integer, x, is 45.
The second integer, x + 1, is 46.
Check your answers.

6) Find three consecutive even Integers whose sum is 156.

Select the variable	let x = first integer
	(x + 2) = second integer
	(x + 4) = third integer
Write the equation	x + (x + 2) + (x + 4) = 156
Solve the equation	3x + 6 = 156
	3x + 6 − 6 = 156 − 6
	3x = 150
	x = 50

The integers are
x = 50
x + 2 = 52
x + 4 = 54 *Check your answers.*

EXERCISES

Solve each algebra word problem using a variable and an equation.

1) Six times a number less 7 is 41. Find the number.

2) Eight more than three times a number is 194. Find the number.

3) The difference between seven times a number and three times that same number is 20. Find the number.

4) Two more than three times a number is eight more than that number. Find the number.

5) One-third a number less three is 12. Find the number.

6) One number is twice the value of another number. Their sum is 96. Find the numbers.

7) Julie and John earned a total of 72 dollars. If Julie earned three times as much as John, how much did each of them earn?

8) In a school, the number of seventh graders is 215 more than the number of eighth graders. If the total of the two grades is 895 students, how many seventh graders and how many eighth graders are there?

9) Steve and Stan sold a total of 93 candy bars for a fund-raiser. If Steve sold 5 more than 3 times the number that Stan sold, find the number of candy bars sold by each.

10) Find two consecutive integers whose sum is 125.

11) Find 3 consecutive integers whose sum is 99.

12) Find 3 consecutive odd integers whose sum is 159.

REVIEW

Solve each of the following.

1) Find 15% of 150.

2) 6 is what % of 125?

3) 3 is 5% of what?

4) Find 150% of 60.

10-3 Time, Rate, and Distance Problems

Solving **time, rate, and distance problems** are easy if you approach them in an organized fashion. You need to know the following formula.

Distance = Rate x Time which is commonly written **D = R x T**

Also, you need to know the two related formulas: $T = \dfrac{D}{R}$ and $R = \dfrac{D}{T}$

Remember the following when solving time, rate, and distance problems.

Helpful Hints
- **First**, read the problem **carefully** and be sure that you fully understand it. Be sure you understand that which is **given**, and what is to be **found**.
- Always draw a sketch. This makes it easier to understand the problem.
- Keep in mind the formula **Distance = Rate x Time**
- Use a chart or table to do your work. Assign a variable to one of the unknowns.
- There are **four** basic types of motion problems
  1. **Separation** problems where two objects start from the same place moving in opposite directions.
  2. **Come-together** problems where two objects start from different places and move towards each other.
  3. **Catch-up** problems where one object leaves a place and a second object leaves the same location at a later time and catches up.
  4. **Back-and-Forth** problems where an object goes out from a place, turns around and comes back using the same route.
- Check your answers.

EXAMPLES

Solve each of the following. Identify the type, draw a sketch, and use a chart.

1) Two cars are 360 km apart. They travel toward each other, one at 80 km per hour and the other at 100 km per hour. How much time will it take before they meet?

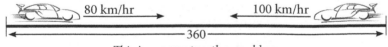

This is a come-together problem.

	rate	x time	= distance
Fast car	100	t	100t
Slow car	80	t	80t

Let t = the number of hours traveled by each car

The total distance is 360

$$100t + 80t = 360$$
$$180t = 360$$
$$t = 2$$

It will take 2 hours before they meet. *Check your answer.*

10-3 Time, Rate, and Distance Problems

2) **Bill and Annie each leave home driving in opposite directions for 3 hours and are then 510 km apart. If Bill's speed was 80 km per hour, what was Annie's speed?**

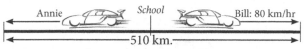

This is a separation problem.

	rate	x time	= distance
Bill	80	3	240
Annie	r	3	3r

Let r = Annie's rate

The total distance is 510 km.

$$3r + 240 = 510$$
$$3r + 240 - 240 = 510 - 240$$
$$3r = 270$$
$$r = 90$$

Annie's speed (rate) was 90 km/hr.

Check your answer.

3) **Mary left her house by bicycle, traveling 40 km per hour. Two hours later, John left the same house in a car trying to catch up with Mary. If he was traveling at a rate of 60 km per hour, how long would it take John to catch up with Mary?**

This is a catch-up problem.

	rate	x time	= distance
Mary	40	t+2	40 (t + 2)
John	60	t	60t

Let t = the number of hours that John will travel

They each traveled the same distance.

$$60t = 40 (t + 2)$$
$$60t = 40t + 80$$
$$60t - 40t = 40t - 40t + 80$$
$$20t = 80$$
$$t = 4$$

John caught up in 4 hours.

Check your answer.

4) Ron rides his bike from home to the lake traveling at a rate of 25 km per hour. He returns home by train at a rate of 75 km per hour. If the bike trip took 2 hours longer than the train trip, how far was it from his home to the lake?

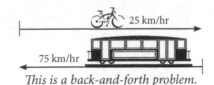

25 km/hr

75 km/hr

This is a back-and-forth problem.

	rate	x	time	=	distance
Bike	25		t+2		25 (t + 2)
Train	75		t		75t

Let t = the number of hours traveled by the train

The distances are equal.

$$75t = 25 (t + 2)$$
$$75t = 25t + 50$$
$$75t - 25t = 25t - 25t + 50$$
$$50t = 50$$
$$t = 1$$

Check: $75t = 75(1) = 75$
$25 (t + 2) = 25(3) = 75$

The distance from home to the lake is 75 km.
Check your answer.

5) A car and a motorcycle start traveling towards each other at the same time from locations 405 km apart. The rate of the car is twice the rate of the motorcycle. In 3 hours they pass each other. Find the rate of each.

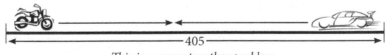

405

This is a come-together problem.

	rate	x	time	=	distance
Motorcycle	r		3		3r
Car	2r		3		6r

The total distance is 405 km.

$$3r + 6r = 405$$
$$9r = 405$$
$$r = 45$$

The rate of the motorcycle is r = 45 km/hr.
The rate of the car is 2r = 90 km/hr.
Check your answers.

EXERCISES

1) Ralph and Joe leave school walking in opposite directions. Ralph walks 1 km per hour faster than Joe. After 2 hours they are 30 km apart. What was the rate of each?

2) Two trains left a station traveling in opposite directions. One traveled at the rate of 60 km per hour, and the other at 72 km per hour. How many hours passed before they were 792 km apart?

3) Two trucks are 180 km apart. They each left at the same time traveling towards each other. One traveled at a rate of 65 km per hour, and the other at 55 km per hour. How many km did each travel before they met?

4) Two planes left the same airport, traveling in opposite directions. One plane traveled 60 km per hour faster than the other. After 5 hours they were 5300 km apart. Find the rate of each.

5) The first runner started a race and maintained a rate of 20 km per hour. One hour later a second runner started the race and maintained a rate of 25 km per hour. How many hours passed before the second runner caught up with the first runner?

6) Steve walked from his house to the lake at a rate of 6 km per hour. He rode a bike back to his house at the rate of 18 km per hour. The walk took 4 hours longer than the bike ride. How far is it from his home to the lake?

7) Elena left her house driving her car at the rate of 45 km per hour. Two hours later, her sister Yana left the house traveling the same direction at a rate of 60 km per hour. How many hours will it take for Yana to catch up with Elena?

8) Sally spent 6 hours walking from her house to the lake and back. She walked to the lake at a rate of 4 km per hour, and walked back to her house at the rate of 2 km per hour. What is the distance from her house to the lake?

REVIEW

Use the quadratic formula $x = \dfrac{-b \pm \sqrt{b^2 - 4ac}}{2a}$ to solve each of the following.

1) $2x^2 + 3x - 9 = 0$ 2) $2x^2 - 5x + 3 = 0$

10-4 Mixture Problems

There are many everyday problems that involve the mixing of products that have different costs. **Mixture problems** have many applications in business, industry, science, and even supermarkets. When working with mixtures it is important to keep in mind the following formula. **Number of items x Price per item = Cost**

It is also very helpful to make a chart when solving mixture problems. Keep the following in mind when solving mixture problems.

Helpful Hints
- **First**, read the problem **carefully** and be sure that you fully understand it. Be sure you understand that which is **given**, and what is to be **found**.
- Use a **chart**.
- Assign a **variable** to one of the unknowns.
- It if often helpful to express some costs as **cents**. Example: $2.75 is 275 cents. It helps avoid having to work with decimals.
- Some mixture problems involve **percents**. It might be good to review the lesson on percents in chapter 1.
- It is often good to use a **calculator** to do the computation.
- Check your answers.

EXAMPLES

Use a chart to solve each of the mixture problems.

1) **A storeowner wants to mix one type of candy worth $2.60 per kg with another type of candy worth $3.60 per kg. He wants 40 kg of a mixture that is worth $3.00 per kg. How many kg of each should he use?**

	# of kg x	price per kg =	cost
$3.60 Candy	n	360	360n
$2.60 candy	40 − n	260	260(40 − n)
Mixture	40	300	40 (300)

Let n = the number of kg of the **$3.60 candy**

Notice that we changed the dollars to cents

The value of the mixture is equal to the value of the $3.60 candy added to the value of the $2.60 candy. Notice the values are changed to cents.

$$360n + 260 (40 - n) = 40(300)$$ *Use a calculator.*
$$360n + 10{,}400 - 260n = 12{,}000$$
$$360n + 10{,}400 - 10{,}400 - 260n = 12{,}000 - 10{,}400$$ *Subtract 10,400 from each side.*
$$360n - 260n = 1{,}600$$
$$100n = 1{,}600$$
$$n = 16$$
$$(40 - n) = 24$$

16 kg of $3.60 candy
24 kg of $2.60 candy
Check your answers.

10-4 Mixture Problems

2) A scientist has a solution which is 15% pure acid. He has a second solution which is 40% pure acid. How many liters of each solution should he use to make 1200 liters of a solution which is 25% pure acid?

	# of liters x	% of acid =	amount of pure acid	
First solution	n	.15	.15n	**Let n = the number of liters of the 15% solution**
Second solution	1200 – n	.4	.4(1200 – n)	
Mixture	1200	.25	.25(1200)	

The sum of the amount of the solutions will equal the amount of the mixture

$$.15n + .4(1200 - n) = .25(1200)$$ *Use a calculator.*

$$.15n + 480 - .4n = 300$$

$$.15n - .4n + 480 - 480 = 300 - 480$$ *Subtract 480 from each side.*

$$-.25n = -180$$

$$.25n = 180$$ n = 720 liters of the first solution

$$n = 720$$ (1200 – n) = 480 liters of the second solution

    Check your answers.

EXERCISES

Use a chart to solve each of the following mixture problems.

1) A company mixes two kinds of cleansers to get a blend selling for 59 cents per liter. One kind is 50 cents per liter, and the other is 80 cents per liter. How much of each should be mixed to get 1,000 liters of blend?

2) Two kinds of candy are mixed to sell at $4.00 per kg. Vanilla candy sells for $3.20 per kg, and chocolate candy sells for $4.40 per kg. How much of each kind is used to make a mixture that weighs 60 kg?

3) A restaurant has a soup which is 24% cream and another soup which is 18% cream. How many liters of each must be used to make 90 liters of soup which is 22% cream?

4) A chemist has one solution which is 30% pure acid and another solution which is 60% pure acid. How many liters of each solution must be used to make 60 liters of solution which is 50% pure acid?

REVIEW

Simplify each of the following.

1) $\dfrac{6x}{5y} \cdot \dfrac{y^2}{2x}$

2) $\dfrac{16x^2}{5} \cdot \dfrac{25}{4y^2}$

3) $\dfrac{6}{y^2 - 9} \div \dfrac{2}{y - 3}$

4) $\dfrac{2x + 2}{5} \div \dfrac{2}{3}$

10-5 Work Problems

INTRODUCTION

Work problems involve the time it takes for people or machines to complete jobs. These problems can apply to such jobs as painting, completing office tasks, construction, manufacturing, and much more. It is good to use a **chart** when completing work problems. Remember the following.

Helpful Hints
- Rate of work x Time = Work done
- **First**, read the problem **carefully** and be sure that you fully understand it. Be sure you understand that which is **given**, and what is to be **found**.
- Use the same units of measure. For example, don't mix days with minutes.
- Fractions are used to show parts of the job.
- Be sure to understand what is being asked for, and answer that question.
- Check your answers.

EXAMPLES

Solve each of the following.

1) **Mary can clean the house in 2 hours. Tom can clean it in 3 hours. If the work together, how long will it take to clean the house?**

	Mary	Tom	Together
Hours needed	2	3	n
Part completed in one hour	$\frac{1}{2}$	$\frac{1}{3}$	$\frac{1}{n}$

Let n = the number of hours needed to complete the job when they work together

$$\begin{array}{ccc} \text{Part Mary does} & & \text{Part Tom does} & & \text{Part of job} \\ \text{in one hour} & + & \text{in one hour} & = & \text{done in one hour} \\ \frac{1}{2} & & \frac{1}{3} & & \frac{1}{n} \end{array}$$

$$\frac{1}{2} + \frac{1}{3} = \frac{1}{n}$$

$$6n \cdot \frac{1}{2} + 6n \cdot \frac{1}{3} = 6n \cdot \frac{1}{n} \quad \textit{Multiply each term by LCM = 6n}$$

$$3n + 2n = 6 \quad \textit{Solve for n}$$

$$5n = \frac{6}{5}$$

$$n = 1\frac{1}{5}$$

From the chart, n = the number of hours to complete the job together = $1\frac{1}{5}$ hours.

---

2) **John can paint a room in 5 hours. If he and Ellen together can paint it in 2 hours, how long would it take Ellen to paint the room alone?**

	John	Ellen	Together
Hours needed	5	x	2
Part completed in one hour	$\frac{1}{5}$	$\frac{1}{x}$	$\frac{1}{2}$

Let x = the number of hours needed by Ellen to complete the job

$$\begin{array}{ccc} \text{Part John does} & & \text{Part Ellen does} & & \text{Part of job} \\ \text{in one hour} & + & \text{in one hour} & = & \text{done in one hour} \\ \frac{1}{5} & & \frac{1}{x} & & \frac{1}{2} \end{array}$$

$$\frac{1}{5} + \frac{1}{x} = \frac{1}{2}$$

$$10x \cdot \frac{1}{5} + 10x \cdot \frac{1}{x} = 10x \cdot \frac{1}{2} \quad \textit{Multiply each term by LCM = 10x}$$

$$2x + 10 = 5x$$

$$5x = 2x + 10 \quad \textit{Solve for x}$$

$$5x - 2x = 2x - 2x + 10$$

$$3x = 10$$

$$x = 3\frac{1}{3}$$

From the chart, x = the number of hours Ellen needs to complete the job alone = $3\frac{1}{3}$ hours.

3) **A pump can fill a tank in 5 hours. Another pump can empty the tank in 6 hours. If both pumps are on how many hours are needed to fill the tank?**

Let x = the number of hours needed to fill the tank with both pumps open.

	Time	Part	Part in hours
Fill	5	$\frac{1}{5}$	$\frac{x}{5}$
Empty	6	$\frac{1}{6}$	$\frac{x}{6}$

Keep in mind that each pump works for x hours. The tank fills faster than it empties.

Amount Filled	–	Amount Emptied	=	Full Pool
$\frac{x}{5}$	–	$\frac{x}{6}$	=	1

$$30 \cdot \frac{x}{5} - 30 \cdot \frac{x}{6} = 30 \cdot 1 \quad \textit{Multiply by LCM = 30}$$
$$6x - 5x = 30$$
$$x = 30 \quad \text{It would take 30 hours to fill the tank.}$$

EXERCISES

Solve each of the following. Use a chart.

1) Julie can clean the windows of a house in 3 hours. Her brother, John, can clean the windows in 6 hours. How long will it take if they work together?

2) Steve can mow a lawn in 30 minutes. Jose can mow the same lawn in 20 minutes. How long will it take if they work together?

3) A father and his son can paint a house in 3 days. Working alone, the father can paint the house in 4 days. How long would it take the son if he worked alone?

4) A pipe can fill a swimming pool in 3 hours. Another pipe can empty the swimming pool in 6 hours. If the pool is empty and both pipes are opened, how long will it take to fill the swimming pool?

5) Susan can paint a room in 6 hours. If her sister helps her, the job is finished in 4 hours. How long would it take Susan's sister to paint the room if she worked alone?

6) Dan and Dave can work together and paint a house in 4 days. Dan works twice as fast as Dave. How long would it take each of them to paint the house alone?

7) Phil can build a fence in 2 hours. Steve can build the fence in 6 hours. Phil painted alone for 1 hour and had to quit working. How long would it take Steve to complete the job?

8) One secretary can type a report in 12 hours. Another secretary can do the job in 18 hours. How long would it take if both secretaries worked together to complete the job?

REVIEW

Solve each of the following.

1) Six less than twice a number is 16. Find the number.

2) Four times a number less than 6 is 8 more than twice the number. Find the number.

10-6 Age Problems

Age problems can be solved with or without a chart. We will approach the age problems without using a chart. You will have to represent **past** and **future** ages when working these problems. To represent a **past** age, simply **subtract** from the present age. To represent a **future** age, simply **add** to the present age. Use the following steps when solving age problems.

Helpful Hints
- **First**, read the problem **carefully** and be sure that you fully understand it. Be sure you understand that which is **given**, and what is to be **found**.

- **Second**, select a variable to represent one of the unknowns. This variable will be used to describe all the other numbers in the problem. Often it is good to have the variable represent the smallest number in the problem.

- **Third**, translate the problem into an equation.

- **Fourth**, solve the equation and use the solution to answer the question that was asked in the problem. Sometimes the answer will be the value of the variable. Sometimes it will be necessary to use the value of the variable to find what was asked for in the problem.

- **Fifth**, check your answers.

EXAMPLES

Solve each of the following.

1) **Steve is 6 years older than Al. If the sum of their ages is 30, find each of their ages.**

Let x = Al's age

x + 6 = Steve's age

$x + (x + 6) = 30$ *Write the equation and solve.*
$2x + 6 = 30$
$2x + 6 - 6 = 30 - 6$
$2x = 24$
$x = 12$

Al's age = x = 12
Steve's age = x + 6 = 18
Check your answers.

2) **Angela is 3 times as old as Linda. In 5 years Angela will be twice as old as Linda will be. Find their present ages.**

Let x = Linda's age

Let 3x = Angela's age

(x + 5) = Linda's age in 5 years
(3x + 5) = Angela's age in 5 years

$3x + 5 = 2(x + 5)$ *Write the equation and solve.*
$3x + 5 = 2x + 10$
$3x - 2x + 5 = 2x - 2x + 10$
$x + 5 = 10$
$x + 5 - 5 = 10 - 5$
$x = 5$

Linda's age = x = 5
Angela's age = 3x = 15
Check your answers.

10-6 Age Problems

3) **Juan is 5 years older than Amir. Five years ago, Juan was twice as old as Amir. Find their present ages.**

Let x = Amir's age

Let x + 5 = Juan's age

$(x + 5) - 5 = 2(x - 5)$ *Write the equation and solve.*

$x = 2x - 10$

$x - x = 2x - x - 10$

$0 = x - 10$

$10 = x$

Amir's age = x = 10

Juan's age = x + 5 = 15

Check your answers.

4) **Elena's mother is 26 year older than Elena. In 10 years the sum of their ages will be 80. What are their ages now?**

Let x = Elena's age

x + 26 = Mother's age

$(x + 10) + (x + 26 + 10) = 80$ *Write the equation and solve.*

$2x + 46 = 80$

$2x + 46 - 46 = 80 - 46$

$2x = 34$

$x = 17$

Elena's age = x = 17

Mother's age = x + 26 = 43

Check your answers.

EXERCISES

Solve each of the following.

1) A mother is 6 times as old as her son. In 6 years, the mother will be 3 times as old as her son. What are their present ages?

2) Moe is 8 years older than Zach. If twenty years ago Moe was three times as old as Zach, what are their present ages?

3) Dave is now twice as old as Chuck. Six years ago, Dave was three times as old as Chuck. Find their present ages.

4) Laura is 25 years older than Larry. In 10 years from now, Laura will be twice as old as Larry will be. Find their present ages.

5) Sam is now 7 years older than Lance. In 20 years, the sum of their ages will be 81. What are their present ages?

6) Vica is twice as old as Carly. Ten years ago, the sum of their ages was 70. How old is each of them now?

7) Sophie is 4 times as old as Rhoda. In three years, Sophie will be three times as old as Rhoda. What are their present ages?

8) Eric is now four times as old as Ralph. Five years ago, Eric was nine times as old as Ralph. How old is each of them now?

9) A father is 40 years older than his son. In 10 years, the sum of their ages will be 80. What are their present ages?

10) Jake is three times as old as Olivia. Four years ago, Jake was 4 times as old as Olivia was at that time. What are their present ages?

REVIEW

Use the quadratic formula $x = \dfrac{-b \pm \sqrt{b^2 - 4ac}}{2a}$ to solve each of the following.

1) $2x^2 + 7x - 4 = 0$

2) $2x^2 - 7x + 2 = 0$

10-7 Coin Problems

When working with **coin problems** it is important to keep the **value** of the coins in mind. For example, three nickels has a value of 3(5) = fifteen cents. When solving coin problems it is usually a good idea to express the values in **cents**. For example, $3.25 would be expressed as 325 cents. Remember the following when solving coin problems.

Helpful Hints

- **First**, read the problem **carefully** and be sure that you fully understand it. Be sure you understand that which is **given**, and what is to be **found**.

- **Second**, select a variable to represent one of the unknowns. This variable will be used to describe all the other numbers in the problem. Often it is good to have the variable represent the smallest number in the problem.

- **Third**, translate the problem into an equation. Remember to keep the **value** of the coins in mind when writing the equation.

- **Fourth**, solve the equation and use the solution to answer the question that was asked in the problem. Sometimes the answer will be the value of the variable. Sometimes it will be necessary to use the value of the variable to find what was asked for in the problem.

- **Fifth**, check your answers.

EXAMPLES

Solve each of the following.

1) **A man has twice as many quarters as nickels. The total value of the coins is $4.40. How many of each coin does he have?**

Let n = the number of nickels

2n = the number of quarters

$5n + 25(2n) = 440$ *Write the equation and solve.*

$5n + 50n = 440$ *Keep the value of each coin in mind*

$55n = 440$

$n = 8$

n = 8 nickels

2n = 16 quarters

Check your answers.

2) **A boy has saved 58 coins consisting of dimes and nickels. The total value of the coins is $4.80. How many of each coin has he saved?**

Let n = the number of nickels

58 – n = the number of dimes

$5n + 10(58 - n) = 480$ *Write the equation and solve.*

$5n + 580 - 10n = 480$ *Keep the value of each coin in mind*

$-5n + 580 = 480$

$-5n + 580 - 580 = 480 - 580$

$-5n = -100$

$\frac{-5n}{-5} = \frac{-100}{-5}$

$n = 20$

n = 20 nickels

58 – n = 38 dimes

Check your answers.

10-7 Coin Problems

3) **A girl has a collection of nickels, dimes, and quarters whose total value is $11.25. She has 3 times as many nickels as dimes, and 5 more quarters than dimes. How many of each coin are in her collection?**

Let x = the number of dimes

3x = the number of nickels

x + 5 = the number of quarters

$10x + 5(3x) + 25(x + 5) = 1125$	*Write the equation and solve.*
$10x + 15x + 25x + 125 = 1125$	*Keep the value of each coin in mind*
$50x + 125 = 1125$	
$50x + 125 - 125 = 1125 - 125$	x = 20 dimes
$50x = 1000$	3x = 60 nickels
$x = 20$	x + 5 = 25 quarters
	Check your answers.

EXERCISES

Solve each of the following.

1) A girl has a collection of dimes and quarters. She has 4 times as many quarters as dimes. If the total value of her collection is $2.20, how many dimes and how many quarters does she have?

2) A man has 60 coins made up of dimes and quarters. If the total value is $9.60, how many of each of the coins does he have?

3) A woman has 7 times as many dimes as nickels. The total value of the coins is $3.75. How many of each coin does she have?

4) A boy has 3 times as many dimes as nickels. The total value is $2.80. How many of each coin does he have?

5) A collection of 40 dimes and quarters has a value of $6.40. How many of each coin type are in the collection?

6) A man has a collection of nickels, dimes, and quarters whose value is $6.55. He has 3 times as many quarters as dimes, and 5 more nickels than dimes. How many of each type of coin does he have?

7) A woman's purse contains nickels and dimes whose total value is $2.70. If there are 30 coins in all, how many are nickels and how many are dimes?

8) A man has $5.75 in nickels, dimes, and quarters. He has 3 times as many nickels as dimes, and 7 more quarters than dimes. How many of each coin does he have?

REVIEW

Solve using the substitution method.

1) $x = y + 3$
   $x + y = 7$

2) $y = 2x - 3$
   $4x + y = 9$

10-8 Investment Problems

INTRODUCTION

Everyone invests money in something. Some of the common financial investments are stocks, bonds, and savings accounts. It is important to know the following formula.

Principal x Rate = Income

Principal represents the amount of money invested. **Rate** represents the annual rate of interest, which is a percent. **Income** represents the annual income, which is sometimes called interest, earnings, or return. It is helpful to use a chart when solving investment problems. Also, using a calculator for computation is helpful. You might want to review the lessons on percent from chapter one. Remember the following when solving investment problems.

Helpful Hints
- **First**, read the problem **carefully** and be sure that you fully understand it. Be sure you understand that which is **given**, and what is to be **found**.
- Assign a variable to one of the unknowns.
- Use a chart.
- Eliminate decimals from equations when possible. Multiplying each side of the equal sign by a power of ten does this.
- Use a calculator when completing the computation.
- Check your answers.

EXAMPLES

Solve each of the following.

1) **Jean invested $20,000, part at 7% and the rest at 5%. The interest earned in one year was $1,280. How much did Jean invest at each rate?**

Let x = the number of dollars invested at 7%

Investment	Principal x	Rate =	Income
7% Investment	x	.07	.07x
5% Investment	$20,000 – x	.05	.05($20,000 – x)

The total annual income = $1,280

$.07x + .05(\$20,000 - x) = \$1,280$ *Write the equation.*

$7x + 5(\$20,000 - x) = \$128,000$ *Multiply both sides of the equal sign by 100 to eliminate the decimals.*

$7x + \$100,000 - 5x = \$128,000$

$7x - 5x + \$100,000 - \$100,000 = \$128,000 - \$100,000$ *Subtract $100,000 from each side*

$2x = \$28,000$

$x = \$14,000$

$x = \$14,000$ @ 7%

$\$20,000 - x = \$6,000$ @ 5% *Check your answers.*

10-8 Investment Problems

2) **Robin made an investment. Part was invested at an interest rate of 6% and the remaining $2,000 was invested at 5%. The income for one year was $460. How much was invested at 6%?**

Let x = the number of dollars invested at 6%

Investment	Principal x	Rate =	Income
6% Investment	x	.06	.06x
5% Investment	$2,000	.05	$100

The total annual income $= \$460$

$.06x + \$100 = \$460$		*Write the equation*
$6x + \$10,000 = \$46,000$		*Mulitply both sides by 100*
$6x = \$36,000$		
$x = \$6,000$		
$x = \$6,000 @ 6\%$		*Check your answer*

EXERCISES

Solve each of the following.

1) Donald invested some money at 6%. He also invested twice that amount at 5%. If the total annual income was $640, how much did he invest at each rate?

2) Susan invested $2,000. Part of it at 6% and the rest at 9%. If the annual income was $138, how much did she invest at each rate?

3) Sherry invested a sum of money at 6% and invested a second sum that was $1,500 greater than the first sum, at 5%. If the total annual income was $570, how much was invested at each rate?

4) A man invested $200,000. Part at 6% and the rest at 7%. The annual income was $13,400. How much was invested at each rate?

5) Sheila invested a certain sum of money at 6%. She also invested a second sum, which was $2,000 less than the first sum, at 5%. Her total annual income was $890. How much did she invest at each rate?

6) Sam invested a sum of money at 6%. He invested a second sum, $1,000 more than the first sum at 7%. If the annual income was $1,110, how much did he invest at 6%? How much at 7%?

REVIEW

Multiply each of the following.

1) $(-6xy)(2x^2 y^2)$

2) $(x^2 y^4)(-3x^3 y^5)$

3) $(r^2 s^4 t^3)(r^2 st^2)$

4) $(5x^2)(4x^4)$

Chapter 10 Review: Word Problems

Solve each of the following.

1) Three times a certain number less 17 is equal to 28. Find the number.

2) One number is four times another. The difference between the two numbers is 48. Find the two numbers.

3) If 12 more than 9 time a number is equal to 10 times that number plus 4, find the number.

4) The sum of two numbers is 88. One number is 24 greater than the other. Find the two numbers.

5) Two cars are 700 km apart. They drive towards each other. One travels at 40 km/hr and the other at 30 km/hr. How many hours before they meet?

6) Two trains start from the same location and travel in opposite directions. One train travels 60 km/hr faster than the other. After 5 hours they are 1,500 km apart. Find the rate of each train.

7) Tom leaves his house traveling at 45 km/hr. Two hours later, his brother, Steve, leaves the house traveling 60 km/hr. How many hours will it take Steve to catch up with Tom?

8) A man drove from his house to the lake at a rate of 60 km/hr. He returned home traveling at a rate of 45 km/hr. The entire trip took 7 hours. How far is it from his house to the lake?

9) A store owner has some coffee worth $1.80 per kg and some worth $2.80 per kg. He wants to make 60 kg of a mixture that will be worth $2.20 per kg. How many kgs of each type of coffee should he use?

10) Steve can mow the lawn in 30 minutes by himself. Cathy can mow the same lawn in 50 minutes by herself. How long will it take to mow the lawn if they work together?

11) Peter and Paul can paint a room in 6 hours if they work together. If Peter works alone, it takes him 10 hours. How many hours will it take Paul to paint the room if he works alone?

12) Steve has a collection of nickels and dimes. He has 8 more nickels than dimes. The value the coins is $2.80. How many coins of each type does he have?

13) Ellen has 40 coins which are all nickels and dimes. If the value of the coins is $3.20, how many of each coin does she have?

14) Maria is 5 years older than Sophia. Five years ago, Maria was two times as old as Sophia. What are their ages now?

15) Sonya is three times as old as Harry. In 18 years from now, Sonya will be twice as old as Harry. What are their present ages?

16) A man invested $9,000. Some was invested at 6% and some at 7%. The earnings for one year was $575. How much was invested at each rate?

Chapter Quiz 1

Solve each of the following:

1) -626 + 342

2) -96 + 13 + -12

3) 16 – 19

4) -96 – 72

5) (-26)(12)

6) $\dfrac{368}{-16}$

Simplify:

7) $-\dfrac{3}{4} + -\dfrac{1}{3}$

8) $-\dfrac{3}{4} \times -1\dfrac{1}{2}$

9) $-2\dfrac{1}{2} \div -1\dfrac{1}{4}$

10) 5.32 x -1.2

Rewrite as an Integer:

11) 3^4

Simplify:

12) $7^3 \bullet 7^5$

13) $(5^3)^3$

14) $\sqrt{121}$

Solve:

15) $7(6 + 7) - 5^2$

16) $3 + \{4(8 - 2) \div 6\} - 5$

Name the property that is illustrated:

17) $1 \times \dfrac{3}{4} = \dfrac{3}{4}$

1)	
2)	
3)	
4)	
5)	
6)	
7)	
8)	
9)	
10)	
11)	
12)	
13)	
14)	
15)	
16)	
17)	

Chapter Quiz 1

Write the numbers in order from least to greatest:

18) $2^3, 5^2, 3^3$

Solve:

19) Is M = {(2,3), (4,5), (7,8), (7,9)} a function? Why?

Use shortcut division to find the prime factorization of each of the following. Express your answers using exponents.

20) 54   21) 200

Find the greatest common factor: 22) 75, 50

Find the least common multiple: 23) 24, 20

Change each of the following to scientific notation:

24) 14,490,000,000   25) .00000019

Solve:

26) $\frac{4}{x} = \frac{12}{60}$

27) A train can travel 90 miles in $1^1/2$ hours. At this rate, how far will the train travel in 6 hours?

28) Find 20% of 210.   29) 18 is what % of 24?

30) There are 50 sixth graders in a school. This is 20% of the school. How many students are in the school total?

18)	
19)	
20)	
21)	
22)	
23)	
24)	
25)	
26)	
27)	
28)	
29)	
30)	
Score:	

Chapter Quiz 2

Solve each equation.

1) $x + 15 = -8$ 2) $x + 2.2 = 7.6$

3) $n + 9 = -13$ 4) $9x = -36$

5) $\frac{-x}{9} = -6$ 6) $-x = 45$

7) $\frac{x}{3} + 4 = 13$ 8) $2x + 2.8 = 9.6$

9) $-2x + 4 = 12$ 10) $5x + 3 = 2x + 15$

11) $3x - 12 = 20 - 2x$ 12) $7x + 3 = 2x + 15$

Use the distributive property to solve each of the following.

13) $-6 (2x + 6) = 48$ 14) $7 (x - 2) + 17 = 3$

15) $5 (x + 3) + 9 = 3 (x - 2) + 6$

16) $5x + 2x - 5 = 30$ 17) $3 (x + 2) - 4x = x + 16$

18) $3x - 2 + 4x + 3 = -5 + 3x + 2x + 16$

19) $|5x + 7| = 32$ 20) $2|y + 1| = 8$

Simplify: 21) $(9x + 3y) - (4x + 12y)$

22) $(-7x^2 + 2y - 3x^2y) + (4x^2 - 5y + 2x^2y)$

Solve: 23) $4 (x + 5) = 3 (x + 2)$

24) $\frac{3x - 2}{2} = 17$ 25) $\frac{x - 5}{4} + 6 = 31$

1) _____
2) _____
3) _____
4) _____
5) _____
6) _____
7) _____
8) _____
9) _____
10) _____
11) _____
12) _____
13) _____
14) _____
15) _____
16) _____
17) _____
18) _____
19) _____
20) _____
21) _____
22) _____
23) _____
24) _____
25) _____

<u>Score:</u> _____

Chapter Quiz 3

Graph each of the following linear equations. Select 4 values for x and find the value for y. Graph and connect the points. Use graph paper.

1) $y = 3x + 1$ 2) $y = \frac{1}{3}x - 1$ 3) $y = \frac{1}{2}x + 4$

Find the x and y intercepts for each equation and draw the graph.

4) $y = 4$ 5) $4x + y = 10$

6) $3x - y = 10$ 7) $2x + 6y = 12$

Find the slope of each line that passes through the given points.

8) $(8, 4), (6, 5)$ 9) $(0, -3), (3, -1)$

10) $(2, 6), (-1, 3)$

Change each of the following linear equations from the standard form to the slope-intercept form and graph the line.

11) $y + 2x = 5$ 12) $6x + 6y = 12$

13) $x - 3y = -6$ 14) $2x - y = 1$

Find the equation of the line in slope-intercept form. Graph the line.

15) Passes through $(2, 5)$ with slope 5. 16) Passes through $(5, 4)$ with slope -5.

Find the equation of a line in the point-slope form for each of the following. $y - y_1 = m (x - x_1)$

17) Slope = 5 passes through $(3, 2)$ 18) Slope = $-\frac{3}{4}$ passes through $(-3, -4)$

1) _____

2) _____

3) _____

4) _____

5) _____

6) _____

7) _____

8) _____

9) _____

10) _____

11) _____

12) _____

13) _____

14) _____

15) _____

16) _____

17) _____

18) _____

Score: _____

Chapter Quiz 4

Graph each of the following inequalities.

1) $x \leq 4$ 2) $x \geq .5$ 3) $x > 10.5$

Solve each inequality. Then graph the solution on a number line.

4) $x + 3 \leq 5$ 5) $x - 3 \leq -7$

6) $x + 6 \geq 16$ 7) $x + 1\frac{1}{2} \leq -3\frac{1}{2}$

8) $-5x \leq -25$ 9) $4x > -12$

10) $\frac{x}{-2} \leq -4$ 11) $\frac{x}{3} > -6$

12) $3x + 4 > 28$ 13) $-6m + 3 > -9$

14) $\frac{2x - 3}{5} \geq 7$ 15) $-3x + 6 \leq -3$

16) $-1 < x + 3 < 5$ 17) $x + 8 < 10 \text{ or } x - 5 > 2$

18) $|x + 5| < 9$ 19) $|2x + 1| \leq 5$

1) _____

2) _____

3) _____

4) _____

5) _____

6) _____

7) _____

8) _____

9) _____

10) _____

11) _____

12) _____

13) _____

14) _____

15) _____

16) _____

17) _____

18) _____

19) _____

Score: _____

Chapter Quiz 5

Use the graphing method to find the ordered pair that is the solution to each system of linear equations. For an accurate graph, use graph paper.

1) $y = 3x$
    $y = -3x + 4$

2) $y = -x + 2$
    $y = -2x + 5$

3) $y = 2x + 7$
    $y = -3x - 3$

Use the substitution method to solve each system of linear equations.

4) $x + y = 6$
    $x = y + 2$

5) $x = y + 2$
    $2y + x = 17$

6) $x = 3y$
    $x + y = 16$

Solve each of the following using the elimination method.

7) $x - 4y = 5$
    $2x - 7y = 9$

8) $x + y = 5$
    $5x - 3y = 17$

Graph each pair of linear inequalities. The solution set of the system is where the shading overlaps. Be sure to use graph paper.

9) $y > 2x + 5$
    $y < -x$

10) $y > x$
    $y > -2x + 4$

1) _____

2) _____

3) _____

4) _____

5) _____

6) _____

7) _____

8) _____

9) _____

10) _____

Score: _____

Chapter Quiz 6

Add each of the following.

1) $\quad y^2 + 2y + 3$
$\quad + \underline{y^2 - 8y + 5}$

2) $(x^2 + 2x + 1) + (3x^2 + 4x - 3)$

Subtract each of the following.

3) $(6x^2 - 5xy + 7y^2) - (3x^2 + 4xy - y^2)$

4) $(-3x^2 - 3xy + 2) - (-4x^2 - 2xy - 4)$

Multiply each of the following.

5) $(3x^2)^3$

6) $(-3y^2)(4x^2y)(2xy^3)$

Divide the following. Express your answer with positive exponents.

7) $\dfrac{54x^2 y^2 z^3}{9xyz^4}$

Multiply the following.

8) $2x^2y$ and $(4xy^2 - 2xy + 3x^2y^2)$

9) $6x^2$ and $(-11x^2 + 3x + 5)$

Multiply the following pairs of binomials.

10) $(x + 6)(x + 7)$

11) $(8x + 3)(5x + 2)$

Divide each of the following.

12) $\dfrac{25x^3 + 15x^2 - 30x}{5x}$

13) $\dfrac{9x^2 y^2 + 3x^2 y - 6xy^2}{-3xy}$

1) _____

2) _____

3) _____

4) _____

5) _____

6) _____

7) _____

8) _____

9) _____

10) _____

11) _____

12) _____

13) _____

Score: _____

Chapter Quiz 6

Find the quotient of each of the following. If there is a remainder, include it in the answer.

14) $(3x^2 + 14x + 15) \div (x + 3)$

15) $(15x^2 - 19x - 56) \div (5x + 7)$

Factor each of the following.

16) $28x^2 + 21x^4$

17) $15x^2y^2 + 25xy + x$

18) $x^2 + 2x - 15$

19) $x^2 + 3x - 54$

20) $25m^2 - 49$

21) $81y^6 - 25y^2$

22) $4x^2 - 6x - 4$

23) $10x^3 - 40x$

24) $3x^2 - 5x - 2$

25) $3x^2 + 7x + 2$

14) _____

15) _____

16) _____

17) _____

18) _____

19) _____

20) _____

21) _____

22) _____

23) _____

24) _____

25) _____

Score: _____

Chapter Quiz 7

Simplify each of the following.

1) $\dfrac{-42x^6y^3}{-7x^4y^4}$

2) $\dfrac{24xyz}{8x^2yz^3}$

3) $\dfrac{3x-3}{x^2-1}$

4) $\dfrac{x^2-2x-15}{x^2-x-12}$

5) $\dfrac{5x^2-5}{5-5x}$

6) $\dfrac{7-x}{x^2-49}$

7) $\dfrac{x+3}{5} = \dfrac{x+1}{4}$

8) $\dfrac{5m+2}{3} = \dfrac{3m-1}{2}$

9) $\dfrac{12a}{5a} \bullet \dfrac{a^2}{6a^2}$

10) $\dfrac{x^2-4}{2} \bullet \dfrac{4}{x-2}$

11) $\dfrac{x}{9} \div \dfrac{x}{3}$

12) $\dfrac{x^2-25}{18} \div \dfrac{x-5}{27}$

13) $\dfrac{5}{2x} + \dfrac{6}{2x}$

14) $\dfrac{b-15}{2b+12} - \dfrac{-3b+8}{2b+12}$

15) $\dfrac{2x}{3} + \dfrac{x}{2}$

16) $\dfrac{3x}{8} - \dfrac{x}{12}$

17) $\dfrac{6x}{x-3} + \dfrac{x}{x+1}$

18) $\dfrac{2x}{2x+5} - \dfrac{5}{2x+5}$

19) $\dfrac{x}{3} - \dfrac{x}{5} = 4$

20) $\dfrac{x}{3} + \dfrac{x}{7} = 10$

1) _____

2) _____

3) _____

4) _____

5) _____

6) _____

7) _____

8) _____

9) _____

10) _____

11) _____

12) _____

13) _____

14) _____

15) _____

16) _____

17) _____

18) _____

19) _____

20) _____

Score: _____

Chapter Quiz 8

Simplify each radical expression.

1) $\sqrt{81x^2}$ 2) $\sqrt{8x^2}$ 3) $\sqrt{x^6}$

4) $\sqrt{75x^2y^2z^2}$ 5) $\sqrt{4x^2y^4z^3}$

Solve each of the following.

6) $\dfrac{x}{9} = \dfrac{4}{x}$ 7) $3\sqrt{4x} = 12$

8) $\sqrt{7x} = 7$ 9) $\sqrt{x-5} + 2 = 3$

Solve each of the following. Express in simplest form.

10) $8\sqrt{5} + 3\sqrt{5}$ 11) $\sqrt{18} + \sqrt{50}$

12) $2\sqrt{3} + \sqrt{12}$ 13) $\sqrt{14} \cdot \sqrt{2}$

14) $2\sqrt{3} \cdot 5\sqrt{27}$ 15) $3\sqrt{2} \cdot 4\sqrt{18}$

16) $\dfrac{\sqrt{27}}{\sqrt{3}}$ 17) $\dfrac{\sqrt{30a^3}}{\sqrt{6a^2}}$

18) $\dfrac{8\sqrt{48}}{4\sqrt{2}}$

1) _____

2) _____

3) _____

4) _____

5) _____

6) _____

7) _____

8) _____

9) _____

10) _____

11) _____

12) _____

13) _____

14) _____

15) _____

16) _____

17) _____

18) _____

Score: _____

Chapter Quiz 8

Rationalize the denominator and simplify completely each of the following.

19) $\dfrac{\sqrt{5}}{\sqrt{3}}$ 20) $\dfrac{25}{\sqrt{20}}$ 21) $\dfrac{\sqrt{13}}{\sqrt{2}}$

Rationalize the denominator and simplify each of the following. Express your answers in their simplest form.

22) $\dfrac{3}{4 + \sqrt{5}}$ 23) $\dfrac{20}{\sqrt{6} + 2}$

Using the Pythagorean Theorem, find the length of the missing side. Use a calculator that has a square root function.

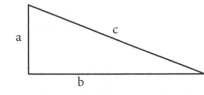

24) $a = 16, b = 63, c = ?$

Draw a sketch and use the Distance Formula to find the distance between the given points.

$$D = \sqrt{(x_2 - x_1)^2 + (y_2 - y_1)^2}$$

25) (-5, 4) (3, -2)

Find the coordinates of the midpoint between the two given points.

$$M = \left(\dfrac{x_1 + x_2}{2}\right), \left(\dfrac{y_1 + y_2}{2}\right)$$

26) (-4, 4) (4, -4)

Chapter Quiz 9

Solve each of the following.

1) $x^2 + 8x + 15 = 0$ 2) $x^2 + 3x = 0$

3) $x^2 + 8x = 0$ 4) $x^2 = 15 - 2x$

5) $x^2 - 144 = 0$ 6) $\dfrac{x}{8} = \dfrac{4}{x}$

7) $7x^2 = 3x^2 + 36$ 8) $2(4x - 2)^2 = 72$

Solve each of the following by completing the square.

9) $x^2 - 4x - 21 = 0$ 10) $x^2 - 6x + 8 = 0$

11) $x^2 + 4x + 3 = 0$

Use the quadratic formula to solve each of the following.

$$x = \dfrac{-b \pm \sqrt{b^2 - 4ac}}{2a}$$

12) $x^2 - 5x + 6 = 0$ 13) $x^2 - x - 5 = 0$

14) $x^2 - 4x - 5 = 0$ 15) $x^2 - 2x - 2 = 0$

Use the 4-step method to graph each of the following.

16) $y = x^2 - 5$ 17) $y = -x^2 + 7$ 18) $y = x^2 - 4x + 3$

Determine how many times the graph of each quadratic function passes through the x-axis by using the discriminant $b^2 - 4ac$.

- If $b^2 - 4ac < 0$, the graph crosses the x-axis 0 times
- If $b^2 - 4ac > 0$, the graph crosses the x-axis 2 times
- If $b^2 - 4ac = 0$, the graph crosses the x-axis 1 time

19) $y = x^2 + 5x - 6$ 20) $y = 3x^2 - 4x + 3$

1) _____

2) _____

3) _____

4) _____

5) _____

6) _____

7) _____

8) _____

9) _____

10) _____

11) _____

12) _____

13) _____

14) _____

15) _____

16) _____

17) _____

18) _____

19) _____

20) _____

Score: _____

Chapter Quiz 10

Translate each of the following into an equation.

1) Twelve more than three times a number is 50.

2) Twice a number increased by 20 equals 48.

3) Fifteen added to five times a number equals three less than seven times the number.

4) Six times the difference of a number and eight is equal to fifteen.

Solve each algebra problem using a variable and an equation.

5) The difference between seven times a number and 23 is 82. Find the number.

6) John is five times as old as Mike. If the sum of their ages is 18, what is the age of each boy?

7) Find two consecutive integers whose sum is 251.

8) One number is 6 time another. If their sum is 70, find the numbers.

Solve each of the following. Identify the type, draw a sketch, and use a chart.

9) Joe and Mary each leave home driving in opposite directions. If Joe travels at the speed of 56 mph, and Mary travels at the speech of 70 mph. In how many hours will they be 504 miles apart?

10) Two cars started from the same point, driving in opposite directions. One traveled at a rate of 40 mph, and the other at a rate of 30 mph. In how many hours would they be 350 miles apart?

11) Susan and Jane are taking a motorcycle trip. Susan gets a 1 hour head start. She is traveling at a rate of 40 mph. Her friend, Jane, leaves to catch up traveling at a rate of 55 mph. How much time will it take Jane to catch up with Susan?

1) _____

2) _____

3) _____

4) _____

5) _____

6) _____

7) _____

8) _____

9) _____

10) _____

11) _____

Score: _____

Chapter Quiz 10

Use a chart to solve each of the following problems.

12) A store sells mixed nuts worth $2.40 per pound and mixed fruit worth $1.50 per pound. How many pounds of each is needed to produce a mixture of 30 pounds to sell for $2.25 per pound?

13) Peanuts sell for $3.50 per pound and cashews sell for $4.75 per pound. How many pounds of each are needed to make 20 pounds of a mixture that will sell for $4.00 per pound?

14) A scientist has 200 liters of a solution that is 40% acid. How many liters of water should be added to make a solution that is 25% acid?

15) John can paint a room in 6 hours. Mary can paint the same room in 9 hours. How long will it take to paint the room if they work together?

16) Jane can type a 50-page paper in 8 hours. If Jane and Mary can type the paper in 6 hours working together, how long would it take Mary to type the paper working alone?

17) John can paint a fence in 12 hours. His son can paint the fence in 24 hours. How long will it take to paint the fence if they work together?

Solve each of the following.

18) John is 3 times as old as Tom. In 18 years from now, John will be twice as old as Tom will be then. Find the present age of both.

19) Jose is 20 years older than Maria. Sixteen years ago, Jose was 3 times as old as Maria was then. Find their present ages.

20) Julie is 25 years older than Kim. Ten years from now, Julie will be twice as old as Kim. What are their present ages?

21) Leigh has $2.05 in quarters and dimes. She has 4 more quarters than dimes. How many of each type of coin does she have?

22) Jim has a jar of nickels and dimes. There are 5 times as many dimes as nickels. If the total value is $4.40, how many nickels are there? How many dimes are there?

Use a chart to solve the following. You may use a calculator.

23) A man invested some money at 6% yearly interest. He also invested $800 at 7%. If the money earned $128 interest in a year, how much was invested at 6%?

12) _____

13) _____

14) _____

15) _____

16) _____

17) _____

18) _____

19) _____

20) _____

21) _____

22) _____

23) _____

Score: _____

Simplify 1–20.

1) $-7 + 6 - 9$

2) $-23 + 16 - 19 - -6$

3) $(-3)(4)(-5)$

4) $\frac{-120}{-15}$

5) $\frac{-45 \div -5}{-12 \div 4}$

6) $-\frac{3}{4} + \frac{1}{3}$

7) $-\frac{1}{2} - \frac{1}{5}$

8) $-1\frac{1}{2} \cdot -\frac{3}{4}$

9) $5\frac{1}{2} \div -\frac{1}{2}$

10) $3.2 + -4.6$

11) $3.76 - 9.3$

12) $5(-3.62)$

13) $\frac{-4.5}{-0.9}$

14) 7^3

15) $5^2 \cdot 5^2$

16) $\frac{\sqrt{144}}{\sqrt{9}}$

17) $63 \div 7 - 3 \times 2 + 4$

18) $9 + \{(4 \times 5) \times 3\}$

19) $4(6 + 2) - 5^2$

20) $\frac{4^2 + 12}{5 + 3(2 + 1)}$

21) Give a prime factorization for 126.

22) Find the greatest common factor of 170 and 120.

23) Find the least common multiple of 24 and 30.

24) Write 321,000,000 in scientific notation.

25) Write .0000271 in scientific notation.

26) Solve the proportion $\frac{8}{5x} = \frac{2}{5}$

27) Find 60% of 240.

28) 80 is 25% of what?

29) 120 is what % of 150?

Solve each of the following.

30) $n - {-6} = -12$

31) $\frac{1}{3}x - 3 = -3$

32) $\frac{x}{5} + 9 = -11$

33) $\frac{2x}{3} = 8$

34) $\frac{1}{2}n = 3\frac{1}{2}$

35) $6n + 7 = 4n + 5$

36) $3x - 8 = 13 - 4x$

37) $3(x - 5) + 19 = 22$

38) $2x - 3x + 5x = 32$

39) $|m - 6| = 8$

40) $|3x| = 12$

41) $\frac{4x - 28}{3} = 2x$

42) $5(n + 2) = 3(n + 6)$

43) $3(x - 2) + x = 2(x + 1)$

44) $7(m + 2) - 4m = 2(m + 10)$

45) Sketch the graph of $y = x + 3$.

46) Sketch the graph of $y = 4x - 2$.

47) Find the x and y intercepts for $2x + y = 3$.

48) Find the slope of the line that passes through the points $(6, 5)$ and $(8, 4)$.

49) Find the slope of the line that passes through the points $(-2, -3)$ and $(0, -1)$.

50) Solve for y to find the slope of the line with the equation $3x + 2y = 6$.

51) Find the equation of the line in slope-intercept form that has
slope = $-\frac{1}{2}$, y-intercept = 6.

52) Find the equation of the line in slope-intercept form with
slope = 2 and passes through $(-5, 1)$

53) Find the equation of the line in slope-intercept form that passes through
the points $(-2, 3)$ and $(0, 4)$.

54) Find the point-slope form of a line with slope = 3 that passes through (1, 2).

55) Write the equation $y = \frac{2}{3}x + 4$ in standard form.

56) Sketch the graph of the inequality $x \leq 3$.

57) Sketch the graph of the inequality $-2 < x < 3$.

58) Sketch the graph of the inequality $x \leq -2$ or $x > 2$.

59) Sketch the graph of the inequality $|x| \leq 3$.

60) Sketch the graph of the inequality $|x| > 4$.

Solve each inequality.

61) $y - 2 < -10$

62) $-5n - 15 \leq 10$

63) $\frac{x}{2} - 4 > 6$

64) $|x + 1| \leq 3$

65) $|x + 5| > 8$

66) $2|x| > 12$

67) $3|x| < 15$

68) Sketch the graph of $y < 2x$.

69) Sketch the graph of $y \geq 4x - 1$.

70) Solve by using the substitution method.
$$y = 2x$$
$$x + y = 21$$

71) Solve by using the substitution method.
$$2x - 9y = 1$$
$$x - 4y = 1$$

72) Solve by using the elimination method.
$$3x + y = 0$$
$$6x - y = 18$$

73) Solve by using the elimination method.
$$5x + 2y = 16$$
$$3x + 4y = 4$$

74) Sketch the graph that shows the solution set of the following system of inequalities.
$$y > 3$$
$$x < -2$$

75) Add $(4x^2y + 4xy^2 - 2y^3 + 3) + (2x^2y + 3y^3 - 4)$

76) Add $(2xy + 2x + 3y) + (4xy - 5x + 5y)$

77) Subtract $(4x^2 - 3xy - 5y^2) - (2x^2 + xy - 3y^2)$

78) Subtract $(3x^2 - 2xy - 2y^2 - 7) - (-x^2 - 5xy + 4y^2 + 2)$

79) Multiply $(-5x^3y^2)(6xy^4)$

80) Multiply $\dfrac{25x^2y}{3} \cdot \dfrac{12x^3y^5}{5}$

81) Divide $\dfrac{-40x^9y^6}{8x^3y^2}$

82) Divide $\dfrac{7x^3y^4z^7}{x^2yz^2}$

83) Multiply $-3x(4x^2 - 3x + 4)$

84) Multiply $3x^2y(2x^2 - 3xy - y^3)$

85) Multiply $(x + 3)(x + 2)$

86) Multiply $(5x + 2)(2x - 3)$

87) Divide $\dfrac{5x^2 - 15x^3y + 10x}{5x}$

88) Divide $\dfrac{6r^2s + 9rs^2 - 18r^3}{3r}$

Final Review

Factor each of the following.

89) $6x^2y - 9xy$

90) $x^3 - 4x^2 + 4x$

91) $x^2 + 16x + 48$

92) $64x^2 - 25y^2$

93) $5a^2 - 5b^2$

94) $25x^2 + 50x - 200$

Simplify each of the following

95) $\dfrac{3x + 3y}{7x + 7y}$

96) $\dfrac{n^2 - 9}{n^2 + 3n}$

97) $\dfrac{x^2 + 6x + 9}{x^2 - 9}$

98) $\dfrac{m^2 - 12m + 20}{m^2 + 4m - 12}$

99) $\dfrac{m - 2}{2 - m}$

100) $\dfrac{2y^2 - 50}{10 - 24}$

101) Solve $\dfrac{2x + 3}{4} = \dfrac{x - 6}{8}$

102) Multiply $\dfrac{3x^2}{4} \cdot \dfrac{20y}{12y^2}$

103) Multiply $\dfrac{x^2 - x - 6}{14} \cdot \dfrac{7}{x - 3}$

Final Review

104) Divide $\dfrac{2x^2}{5y^2} \div \dfrac{4x^3}{10y}$

105) Divide $\dfrac{x^2 - x - 6}{x^2 - 9} \div \dfrac{x + 2}{x - 3}$

106) Add $\dfrac{x + y}{3xy} + \dfrac{2x - y}{3xy}$

107) Subtract $\dfrac{2x + 1}{7x} - \dfrac{4x - 3}{7x}$

108) Add $\dfrac{3}{2x} + \dfrac{1}{x}$

109) Add $\dfrac{2}{3x} + \dfrac{1}{6xy}$

110) Add $\dfrac{x + 2}{3} + \dfrac{x + 1}{5}$

111) Subtract $\dfrac{4x}{3} - \dfrac{x}{6}$

112) Subtract $\dfrac{x - 2}{4} - \dfrac{x}{12}$

113) Subtract $\dfrac{3}{2x} - \dfrac{1}{8x^2}$

114) Solve $\dfrac{2}{3} - \dfrac{1}{6} = \dfrac{x}{4}$

115) Solve $\frac{3}{x} + \frac{1}{x} = \frac{7}{8}$

116) Solve $\frac{2}{x} - \frac{4}{3x} = \frac{2}{9}$

Simpify each of the following.

117) $\sqrt{63}$

118) $\sqrt{28x^2}$

119) $\sqrt{75x^3}$

120) $\sqrt{\frac{3x^2}{16}}$

For 121 - 123, rationalize the denominator.

121) $\frac{\sqrt{5}}{\sqrt{7}}$

122) $\sqrt{\frac{7}{8}}$

123) $\frac{9\sqrt{3}}{\sqrt{24}}$

Simpify each of the following.

124) $8\sqrt{5} + 5\sqrt{5}$

125) $\sqrt{18} + \sqrt{2}$

126) $2\sqrt{8} - \sqrt{2}$

127) $\sqrt{2} \cdot \sqrt{8}$

128) $3\sqrt{x^3} \cdot 2\sqrt{x}$

129) Solve $\frac{x}{9} = \frac{4}{x}$

130) Solve $x^2 + 6 = 31$

131) Solve $x^2 + 7 = 52$

132) Rationalize the denominator. $\frac{\sqrt{1}}{\sqrt{3}+1}$

133) Rationalize the denominator. $\frac{\sqrt{6}}{1-\sqrt{6}}$

134) Solve for c.

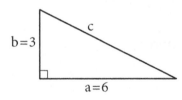

135) Solve for b.

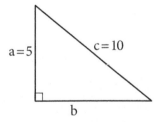

136) Use the Distance Formula to find the distance between the two points (3, -2) and (-5, 4).

137) Use the Midpoint Formula to find the midpoint between the two points (3, 2) and (7, 8).

138) Solve $\qquad$ $3x^2 = 108$

139) Solve $\qquad$ $4x^2 - 7 = 29$

140) Solve $\qquad$ $x^2 + 6x + 9 = 16$

141) Solve $\qquad$ $x^2 - 9x + 8 = 0$

142) Solve $\qquad$ $x^2 - 13x + 40 = 0$

143) Solve $\qquad$ $(x - 2)(x + 5) = -12$

144) Solve $\qquad$ $(x - 3)^2 = 64$

145) Solve by completing the square. $\qquad$ $x^2 - 2x = 5$

146) Solve by completing the square. $\qquad$ $x^2 - 4x = 10$

147) Solve by using the Quadratic Formula. $\qquad$ $2x^2 + 9x - 5 = 0$

148) Solve by using the Quadratic Formula. $\qquad$ $x^2 - 6x + 2 = 0$

149) Use the 4-step process to sketch a graph of the following.
$$y = x^2 - 2x - 3$$

150) Use the 4-step process to sketch a graph of the following.
$$y = x^2 - 4x + 3$$

151) Four times a number less 6 is 8 more than twice that number. Find the number.

152) Rick can ride his bike 2 km/hr faster than Vica. They travel from home in opposite directions for 2 hours. They are then 84 km apart. How far does each of them travel?

153) A company mixes two kinds of cleansers to get a blend they sell for 59 cents a liter. One of the cleansers costs 50 cents per liter, and the other 80 cents per liter. How much of each kind should be mixed to get 2,000 liters of the blend?

154) Allen and his dad want to paint a room. Allen can paint the room by himself in 10 hours. His dad can paint the room by himself in 5 hours. How long will it take if they work together?

155) Susan invested some money at 6% annual interest and 800 dollars at 7%. The money earned 128 dollars interest in a year. How much was invested at 6%?

156) Stuart is now 6 years older than Stan. Six years ago, Stuart was twice as old as Stan. How old is each of them now?

157) Robert has a collection of nickels, dimes, and quarters. The total value of the coins is $8.50. He has 3 times as many nickels as dimes, and 10 more quarters than dimes. How many of each coin does he have?

Glossary

Abscissa The first number in an ordered pair that is assigned to a point on a coordinate plane. Also called the x-coordinate.

Absolute value The distance between 0 and a number on the number line. The absolute value of n is written $|n|$.

Algebra The branch of mathematics that uses letters and numbers to show the relationships between quantities.

Algebraic expression A mathematical expression which contains at least one variable. $2x$, $7x + 9$, $4ab$, and $\frac{x+5}{2}$ are all algebraic expressions.

Associative properties
Addition: $(a + b) + c = a + (b + c)$
Multiplication: $(ab)c = a(bc)$

Axiom A property assumed to be true without proof. Also called a postulate.

Axis of symmetry A line that divides a parabola into two matching parts.

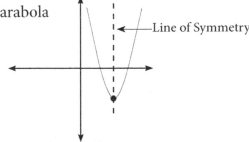

Base The number being multiplied. In an expression such as 4^2, 4 is the base.

Binomial A polynomial with two terms.
$2x + 3y$ and $3x - 2y$ are binomials.

Cartesian coordinate system A system of graphing ordered pairs on a coordinate plane.

Coefficient A number that multiplies the variable. In the term $7x$, 7 is the coefficient of x.

Commutative properties
Addition: $a + b = b + a$
Multiplication: $ab = ba$

Glossary

Completing the square A method for changing a quadratic expression into a perfect square trinomial.

For a quadratic expression in the form $x^2 + bx = c$, follow these steps:
1) Take half of b which is the coefficient of x
2) Square it.
3) Add the result to both sides of the equal sign.
4) The result will be a perfect square trinomial.

Compound inequality Two or more inequalities that are combined using the word "and" or the word "or."

Constant Specific numbers that do not change.

Coordinates An ordered pair of real numbers, which correspond to a point on a coordinate plane.

Coordinate plane The plane which contains the x and y-axes. It is divided into four quadrants.

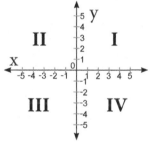

Degree of a polynomial The highest power of a variable that appears in a polynomial expression. The degree of $6x^3 - 4x + 1$ is 3.

Difference of two squares In the difference of two squares, $a^2 - b^2 = (a + b)(a - b)$.

Discriminant The value of $b^2 - 4ac$ is called the discriminant of the quadratic equation $ax + bx + c = 0$. It allows you to determine which quadratic equations have solutions and which ones do not.

Disjoint sets Sets which have no members in common.
(1, 2, 3) and (4, 5, 6) are disjoint sets.

Distance formula The distance D, between the points (x_1, y_1) and (x_2, y_2) on a coordinate plane is:

$$D = \sqrt{(x_2 - x_1)^2 + (y_2 - y_1)^2}$$

Glossary

Distributive Property For real numbers a, b, and c, **a(b + c) = ab + ac.**

Domain of a function The set of all first coordinates (x-values) of the ordered pairs that form the function.

Element of a set Member of a set.

Elimination Method A method of solving a system of linear equations using the following steps:
1) Put the variables on one side of the equal sign and the constant on the other, with the like terms lined up.
2) Add or subtract the equations to eliminate one of the variables. Sometimes it is necessary to multiply one of the equations by a constant first.
3) Solve the equation. Substitute the answer into either of the equations to get the value of the second variable.
4) Check by substituting the answers into the original equations.

Empty set The set that has no members. Also called the null set and is written $\emptyset$ or { }.

Equation A mathematical sentence that contains an equal sign (=) , and states that one expression is equal to another.

Equivalent expressions Expressions which represent the same number.

Evaluate an expression Finding the number an expression stands for by replacing each variable with its numerical value and the simplifying.

Exponent A number that indicates the number of times a given base is used as a factor. In the expression x^2, 2 is the exponent.

Extremes of a proportion In the proportion $\frac{a}{b}=\frac{c}{d}$, a and d are the extremes.

Factor A number or expression that is multiplied to get another number or expression. In the example 4 x 3 = 12, 4 and 3 are factors.

Formula An equation that states a relationship among quantities which are represented by variables. For example, the formula for the area of a rectangle is $A = l \times w$, where A = area, l = length, and w = width.

Function A set of ordered pairs which pairs each x-value with one and only one y-value. For example, F = {(0,2), (-1,6), (4,-2), (-3,4)} is a function.

Glossary

Function notation A way to describe a function that is defined by an equation.
In function notation the equation $y = 4x - 8$ is written as $f(x) = 4x - 8$, where $f(x)$ is read as "f of x" or "the value of f at x."

Graph To show the points named by numbers or ordered pairs on a number line or coordinate plane.

Greatest common factor The largest factor of two or more numbers or terms. Also written GCF.
The GCF of 15 and 10 is 5, since 5 is the largest number that divides evenly into both 10 and 15.
The GCF of 8ab and 6ab is 2ab.

Grouping symbols Symbols used to group mathematical expressions.
Examples include parentheses ( ), brackets [ ], braces { }, and fraction bars —.

Hypotenuse The side opposite the right angle in a right triangle.

Identity Properties
Addition: **a + 0 = 0 + a**
Multiplication: **1 x a = a**

Inequality A mathematical sentence that states that one expression is greater than or less than another. Inequality symbols are read as follows:
< is less than, ≤ is less than or equal to, > is greater than, ≥ is greater than or equal to

Integers Numbers in the set ...-3, -2, -1, 0, 1, 2, 3,... .

Intercept In the equation of a line, the y-intercept is the value of y when x is 0.
The x-intercept is the value of x when y is 0.

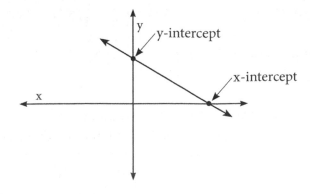

Glossary

Intersection of sets The intersection of two or more sets consists of the members included in all of them. A intersection B is written A∩B.
If set A = {1, 2, 3, 4} and set B = {1, 3, 5} then the intersection sets A and B would be the set {1, 3}.

Inverse operations Operations that "undo" each other. Addition and subtraction are inverse operations. Multiplication and division are inverse operations.

Irrational numbers A real number that cannot be written as the ratio of two integers. They are often represented by non-terminating, non-repeating decimals.
$\sqrt{2} = 1.4142135\ldots$ is an example of an irrational number.

Least Common Denominator (LCD) The least common multiple of the denominators of two or more fractions.
The LCD of $\frac{1}{8}$ and $\frac{1}{6}$ would be 24.
The LCD of $\frac{1}{6x^2}$ and $\frac{1}{4x}$ would be $12x^2$.

Least common multiple (LCM) The least common multiple of two or more expressions is the simplest expression that they will all divide into evenly. To do so, first find the LCM of the coefficients and then the highest degree of each variable and expression.
The LCM of 10x and $25x^2y$ would be $50x^2y$.
The LCM of $4(x - 2)^2$ and $(x + 2)^2 (x - 2)$ would be $4(x - 2)^2 (x + 2)^2$.

Like Terms Terms that have the same variables raised to the same power.
$3xy^2$ and $9xy^2$ are like terms. The coefficients do not have to be the same.

Linear Equation An equation that can be written in the form **Ax + By = C**, where A and B are not both zero. The graph of a linear equation is a straight line.

Means of proportion In the proportion $\frac{a}{b} = \frac{c}{d}$, b and c are the means.

Midpoint formula The midpoint between the two points (x_1, y_1) and (x_2, y_2) is
$$M = \left(\frac{x_1 + x_2}{2}, \ \frac{y_1 + y_2}{2} \right)$$

Monomial A term that is a number, a variable, or the product of a number and one or more variables.
5, x, 4xy, 6xy are all examples of monomials.

Glossary

Multiple A multiple of a number is that number multiplied by an integer.
32 is a multiple of 4 since 4 x 8 = 32. Also, 4 and 8 are factors of 32.

Multiplicative inverse Two numbers whose product is one. They are also called reciprocals.
4 and $\frac{1}{4}$ and $-\frac{2}{3}$ and $-\frac{3}{2}$ are examples of multiplicative inverses.

Natural numbers Numbers in the set 1, 2, 3,… . Also called counting numbers.

Negative exponent For any non-zero number x, and any integer n,
$$x^{-n} = \frac{1}{x^n} \quad \text{and} \quad \frac{1}{x^{-n}} = x^n$$

Negative number A number that is less than zero.
-5 and -3.45 are examples of negative numbers.

Negative slope When the graph of a line slopes down from left to right.

Null set The set that has no members. Also called the empty set which is written ∅ and { }.

Number line A line that represents all real numbers with points.

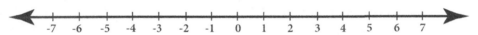

Open sentence A mathematical statement that contains at least one variable.
3x – 6 = 12   n > 25   36 = x – 3   are all open sentences.

Ordered pair A pair of numbers (x, y) that represent a point on the coordinate plane. The first number is the x-coordinate and the second is the y-coordinate.

Order of operations The order of steps to be used when simplifying expressions.
1. Inside the grouping symbols.
2. Exponents
3. Multiply and divide in order from left to right.
4. Add and subtract in order from left to right.

Ordinate The second coordinate of an ordered pair. Also called the y-coordinate.

Origin The point where the x-axis and the y-axis intersect in a coordinate plane.
Written as (0, 0).

Glossary

Parabola The U-shaped curve that is the graph of a quadratic function.

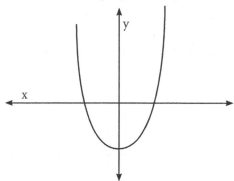

Parallel lines Lines in the same plane that do not intersect. Parallel lines have equal slopes.

Percent Part per hundred or hundredths. Written %.

Perfect square A number than can be expressed as the square of a rational number. The number 36 is a perfect square because it is the square of either 6 or -6.

Perfect square trinomial A trinomial that results from squaring a binomial. Written in the forms $a^2 + 2ab + b = (a + b)^2$ and $a^2 - 2ab + b = (a - b)^2$.

Perpendicular lines Lines in the same plane that intersect at a right (90°) angle. Perpendicular lines have slopes which are negative reciprocals of each other (the product of their slopes is -1).

Point-slope form An equation of a line in the form $y - y_1 = m(x - x_1)$ where m is the slope and (x_1, y_1) is a given point that lies on the line.

Polynomial An algebraic expression of one or more terms connected by plus (+) and minus (–) signs. A monomial has one term. A binomial has two terms. A trinomial has three terms.
3x is a monomial   3 + 5y is a binomial $x^2 + 4x + 3$ is a trinomial

Positive number A number that is greater than 0.
5 and 3.25 are examples of positive numbers.

Positive Slope When the graph of a line slopes up from left to right.

Postulate A property assumed to be true without proof. Also called an axiom

Power An expression that contains a base and an exponent. In the expression x^3, x is the base, and 3 is the exponent.

Glossary

Prime number A prime number is any whole number greater than 1, whose only factors are one and itself.

Prime factorization A whole number that is expressed as a product of its prime factors. The prime factorization of $100 = 2^2 \times 5^2$.

Proportion An equation that states that two ratios are equal.

Pythagorean theorem In a right triangle, if c is the hypotenuse and a and b are the other two legs, then $\mathbf{a^2 + b^2 = c^2}$.

Quadratic equation An equation that can be written in the form $\mathbf{ax^2 + bx + c = 0}$, where $a \neq 0$.

Quadratic formula The formula that can be used to solve any quadratic equation.

$$x = \frac{-b \pm \sqrt{b^2 - 4ac}}{2a}$$

Quadratic function A function that can be written in the form $\mathbf{y = ax^2 + bx + c}$, where $a \neq 0$.

Radical An expression that is written with a radical sign. The expressions $\sqrt{2}$, $\sqrt{x^2}$, and $\sqrt{25}$ are all radicals.

Radicand The expression that is inside the radical sign.

Range of a function The set of all second coordinates (y-values) of the ordered pairs that form the function.

Ratio A comparison of two numbers using division. Written a:b, a to b, and $\frac{a}{b}$.

Rational Expression A fraction whose numerator and denominator are polynomials. $\frac{x+3}{x-2}$ $\frac{x^2-6x+9}{x-3}$ $\frac{32xy^2}{8xy^2}$ are examples.

Rational numbers A number that can be expressed as the quotient of two integers. The denominator cannot equal zero.

Rationalizing the denominator Changing a fraction that has an irrational denominator to an equivalent fraction that has a rational denominator.

Glossary

Real numbers All positive and negative numbers and zero. This includes fractions and decimals.

Reciprocal The multiplicative inverse of a number. Their product is 1.
The reciprocal of 2 is $\frac{1}{2}$. The reciprocal of $-\frac{2}{3}$ is $-\frac{3}{2}$.

Relation Any set of ordered pairs.

Roots The solutions of a quadratic equation.

Rise The change in **y** going from one point to another on a coordinate plane. The vertical change.

Run The change in **x** going from one point to another on a coordinate plane. The horizontal change.

Scientific notation A number written as the product of a number between 1 and 10, and a power of ten. In scientific notation, $7,200 = 7.2 \times 10^3$.

Set A well-defined collection of objects.

Simplified expression The form of an expression with all like terms combined and written in its simplest form.

Slope The steepness of a line. The ratio of the rise (the change in the y direction) to the run (the change in the x direction).
For (x_1, y_1) and (x_2, y_2) which are any two points on a line, **slope** $= \frac{y_2 - y_1}{x_2 - x_1}$, $(x_2 \neq x_1)$

Slope-intercept form An equation of a line in the form $y = mx + b$. The slope is m. The y-intercept is b.

Solution of an equation A number than can be substituted for the variable in an equation to make the equation true.

Solution of an equation containing two variables An ordered pair (x, y) that makes the equation a true statement.

Square Root If $a^2 = b$, then a is a square root of b. Square roots are written with a radical sign $\sqrt{}$.

Glossary

Standard form of a linear equation $Ax + By = C$, where A and B are not both zero.

Standard form of a quadratic equation $ax^2 + bx + c = 0$, where $a \neq 0$.

Subset If A and B are sets and all the members of set A are members of set B, then set A is a subset of set B.

Substitution method A method of solving a system of linear equations using the following steps:
1. Solve one of the equations for either x or y.
2. Substitute the expression from step 1 into the other equation and solve it for the other variable.
3. Take the value from step 2 and substitute it into either one of the original equations and solve it. You will now have two solutions.
4. Check the solutions in each of the original equations.

System of linear equations Two or more linear equations with the same variables.

System of linear inequalities Two or more linear inequalities in the same variable.

Terms Parts of an expression that are separated by addition or subtraction. Terms can be a number, a variable or a product or quotient of numbers and variables.

7, 3x, 4xy, and -3xy are all examples of terms.

Theorem A statement that can be proven to be true.

Transforming an equation To change an equation into an equivalent equation.

Trinomial A polynomial with three terms.
$x^2 + 2xy + y$ is an example of a trinomial.

Undefined rational expression A rational expression that has zero as a denominator. It is meaningless and is considered undefined.

Union of sets If A and B are sets, the union of A and B is the set whose members are found in set A, or set B, or both set A and set B. A union B is written $A \cup B$.

Variable A letter that represents a number

Variable expression Any expression that contains a variable.

Glossary

Venn diagram A type of diagram which shows how certain sets are related.

Vertex of a parabola The maximum or minimum point of a parabola

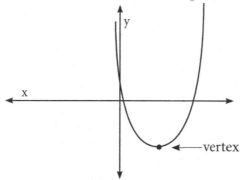

Vertical Line Test A way to tell whether a graph is a function. If a vertical line intersects a graph in more than one point, then the graph is **not** a function.

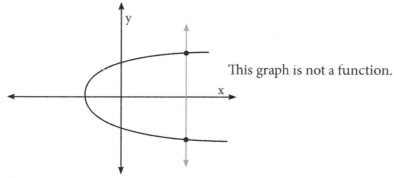

This graph is not a function.

Whole numbers Numbers in the set 0, 1, 2, 3,…

x-coordinate The first number in an ordered pair. Also called the abscissa.

x and y intercepts The points at which a graph intersects the x and y axes.

y-coordinate The second number in an ordered pair. Also called the ordinate.

Zero slope A **horizontal** line has zero slope. The slope of a **vertical** line is undefined.

Zero-product property If the product of two numbers is zero, then at least one of the numbers must equal zero.

Important Formulas

Algebraic Formulas	
Slope formula	$m = \frac{y_2 - y_1}{x_2 - x_1}$
Point-slope form	$y - y_1 = m(x - x_1)$
Slope-intercept form	$y = mx + b$
Standard form of a linear equation	$Ax + By = C$
Quadratic formula	$x = \frac{-b \pm \sqrt{b^2 - 4ac}}{2a}$
Pythagorean theorem	$a^2 + b^2 = c^2$
Distance formula	$D = \sqrt{(x_2 - x_1)^2 + (y_2 - y_1)^2}$
Midpoint formula	$M = (\frac{x_1 + x_2}{2}, \frac{y_1 + y_2}{2})$
Other Formulas	
Average speed	$r = \frac{d}{t}$
Interest	$I = p \times r \times t$
Geometric Formulas	
Perimeter of a polygon	P = the sum of the lengths of the sides
Circumference of a circle	$A = \Pi d$
Area of a rectangle	$A = lw$
Area of a square	$A = s^2$
Area of a parallelogram	$A = bh$
Area of a triangle	$A = \frac{1}{2}bh$
Area of a trapezoid	$A = \frac{1}{2}h(b_1 + b_2)$
Area of a circle	$A = \Pi r^2$
Volume of a cube	$V = s^3$
Volume of a rectangular prism	$V = lwh$

Important Symbols

<	less than		π	pi
$\leq$	less than or equal to		{ }	set
>	greater than		||	absolute value
$\geq$	greater than or equal to		$.\bar{n}$	repeating decimal symbol
=	equal to		$1/a$	the reciprocal of a number
$\neq$	not equal to		%	percent
$\cong$	congruent to		(x,y)	ordered pair
$\approx$	is approximately equal to		f(x)	f of x, the value of f at x
( )	parenthesis		$\perp$	perpendicular
[ ]	brackets		||	parallel to
{ }	braces		$\angle$	angle
...	and so on		$\in$	element of
• or ×	multiply		$\notin$	not an element of
∞	infinity		$\cap$	intersection
a^n	the nth power of a number		$\cup$	union
$\sqrt{}$	square root		$\subset$	subset of
Ø, { }	the empty set or null set		$\not\subset$	not a subset of
$\therefore$	therefore		$\triangle$	triangle
°	degree			

Multiplication Table

x	2	3	4	5	6	7	8	9	10	11	12
2	4	6	8	10	12	14	16	18	20	22	24
3	6	9	12	15	18	21	24	27	30	33	36
4	8	12	16	20	24	28	32	36	40	44	48
5	10	15	20	25	30	35	40	45	50	55	60
6	12	18	24	30	36	42	48	54	60	66	72
7	14	21	28	35	42	49	56	63	70	77	84
8	16	24	32	40	48	56	64	72	80	88	96
9	18	27	36	45	54	63	72	81	90	99	108
10	20	30	40	50	60	70	80	90	100	110	120
11	22	33	44	55	66	77	88	99	110	121	132
12	24	36	48	60	72	84	96	108	120	132	144

Commonly Used Prime Numbers

2	3	5	7	11	13	17	19	23	29
31	37	41	43	47	53	59	61	67	71
73	79	83	89	97	101	103	107	109	113
127	131	137	139	149	151	157	163	167	173
179	181	191	193	197	199	211	223	227	229
233	239	241	251	257	263	269	271	277	281
283	293	307	311	313	317	331	337	347	349
353	359	367	373	379	383	389	397	401	409
419	421	431	433	439	443	449	547	461	463
467	479	487	491	499	503	509	521	523	541
547	557	563	569	571	577	587	593	599	601
607	613	617	619	631	641	643	647	653	659
661	673	677	683	691	701	709	719	727	733
739	743	751	757	761	769	773	787	797	809
811	821	823	827	829	839	853	857	859	863
877	881	883	887	907	911	919	929	937	941
947	953	967	971	977	983	991	997	1009	1013

Squares and Square Roots

No.	Square	Square Root	No.	Square	Square Root	No.	Square	Square Root
1	1	1.000	51	2,601	7.141	101	10201	10.050
2	4	1.414	52	2,704	7.211	102	10,404	10.100
3	9	1.732	53	2,809	7.280	103	10,609	10.149
4	16	2.000	54	2,916	7.348	104	10,816	10.198
5	25	2.236	55	3,025	7.416	105	11,025	10.247
6	36	2.449	56	3,136	7.483	106	11,236	10.296
7	49	2.646	57	3,249	7.550	107	11,449	10.344
8	64	2.828	58	3,364	7.616	108	11,664	10.392
9	81	3.000	59	3,481	7.681	109	11,881	10.440
10	100	3.162	60	3,600	7.746	110	12,100	10.488
11	121	3.317	61	3,721	7.810	111	12,321	10.536
12	144	3.464	62	3,844	7.874	112	12,544	10.583
13	169	3.606	63	3,969	7.937	113	12,769	10.630
14	196	3.742	64	4,096	8.000	114	12,996	10.677
15	225	3.873	65	4,225	8.062	115	13,225	10.724
16	256	4.000	66	4,356	8.124	116	13,456	10.770
17	289	4.123	67	4,489	8.185	117	13,689	10.817
18	324	4.243	68	4,624	8.246	118	13,924	10.863
19	361	4.359	69	4,761	8.307	119	14,161	10.909
20	400	4.472	70	4,900	8.367	120	14,400	10.954
21	441	4.583	71	5,041	8.426	121	14,641	11.000
22	484	4.690	72	5,184	8.485	122	14,884	11.045
23	529	4.796	73	5,329	8.544	123	15,129	11.091
24	576	4.899	74	5,476	8.602	124	15,376	11.136
25	625	5.000	75	5,625	8.660	125	15,625	11.180
26	676	5.099	76	5,776	8.718	126	15,876	11.225
27	729	5.196	77	5,929	8.775	127	16,129	11.269
28	784	5.292	78	6,084	8.832	128	16,384	11.314
29	841	5.385	79	6,241	8.888	129	16,641	11.358
30	900	5.477	80	6,400	8.944	130	16,900	11.402
31	961	5.568	81	6,561	9.000	131	17,161	11.446
32	1,024	5.657	82	6,724	9.055	132	17,424	11.489
33	1,089	5.745	83	6,889	9.110	133	17,689	11.533
34	1,156	5.831	84	7,056	9.165	134	17,956	11.576
35	1,225	5.916	85	7,225	9.220	135	18,225	11.619
36	1,296	6.000	86	7,396	9.274	136	18,496	11.662
37	1,369	6.083	87	7,569	9.327	137	18,769	11.705
38	1,444	6.164	88	7,744	9.381	138	19,044	11.747
39	1,521	6.245	89	7,921	9.434	139	19,321	11.790
40	1,600	6.325	90	8,100	9.487	140	19,600	11.832
41	1,681	6.403	91	8,281	9.539	141	19,881	11.874
42	1,764	6.481	92	8,464	9.592	142	20,164	11.916
43	1,849	6.557	93	8,649	9.644	143	20,449	11.958
44	1,936	6.633	94	8,836	9.695	144	20,736	12.000
45	2,025	6.708	95	9,025	9.747	145	21,025	12.042
46	2,116	6.782	96	9,216	9.798	146	21,316	12.083
47	2,209	6.856	97	9,409	9.849	147	21,609	12.124
48	2,304	6.928	98	9,604	9.899	148	21,904	12.166
49	2,401	7.000	99	9,801	9.950	149	22,201	12.207
50	2,500	7.071	100	10,000	10.000	150	22,500	12.247

Fraction/Decimal Equivalents

Fraction	Decimal	Fraction	Decimal
$\frac{1}{2}$	0.5	$\frac{5}{10}$	0.5
$\frac{1}{3}$	$0.\overline{3}$	$\frac{6}{10}$	0.6
$\frac{2}{3}$	$0.\overline{6}$	$\frac{7}{10}$	0.7
$\frac{1}{4}$	0.25	$\frac{8}{10}$	0.8
$\frac{2}{4}$	0.5	$\frac{9}{10}$	0.9
$\frac{3}{4}$	0.75	$\frac{1}{16}$	0.0625
$\frac{1}{5}$	0.2	$\frac{2}{16}$	0.125
$\frac{2}{5}$	0.4	$\frac{3}{16}$	0.1875
$\frac{3}{5}$	0.6	$\frac{4}{16}$	0.25
$\frac{4}{5}$	0.8	$\frac{5}{16}$	0.3125
$\frac{1}{8}$	0.125	$\frac{6}{16}$	0.375
$\frac{2}{8}$	0.25	$\frac{7}{16}$	0.4375
$\frac{3}{8}$	0.375	$\frac{8}{16}$	0.5
$\frac{4}{8}$	0.5	$\frac{9}{16}$	0.5625
$\frac{5}{8}$	0.625	$\frac{10}{16}$	0.625
$\frac{6}{8}$	0.75	$\frac{11}{16}$	0.6875
$\frac{7}{8}$	0.875	$\frac{12}{16}$	0.75
$\frac{1}{10}$	0.1	$\frac{13}{16}$	0.8125
$\frac{2}{10}$	0.2	$\frac{14}{16}$	0.875
$\frac{3}{10}$	0.3	$\frac{15}{16}$	0.9375
$\frac{4}{10}$	0.4		

Solutions

Lesson 1-1
Page 11

Exercises
1) 14
2) -18
3) -14
4) 37
5) -168
6) -13
7) -4
8) -2
9) -25
10) -138
11) 26
12) -78
13) -110
14) -321
15) -856

Lesson 1-2
Page 12

Exercises
1) -14
2) -3
3) -3
4) 12
5) 9
6) -28
7) 46
8) -49
9) -10
10) -11
11) -103
12) 15
13) -28
14) 9
15) 30
16) -57
17) 51
18) -226

Review
1) -12
2) 7
3) -49
4) -11

Lesson 1-3
Page 13

Exercises
1) -48
2) -126
3) 68
4) -64
5) 288
6) -368
7) -736
8) -24
9) -192
10) 288
11) 72
12) 330

Review
1) -32
2) 36
3) -9
4) -28

Lesson 1-4
Page 14

Exercises
1) -4
2) 6
3) -16
4) 48
5) 15
6) -26
7) -3
8) -2
9) 1
10) -1
11) 3
12) 2

Review
1) 3
2) 30
3) 42
4) -103

Lesson 1-5
Page 15

Exercises
1) $\frac{3}{10}$
2) $\frac{1}{10}$
3) $-\frac{1}{4}$
4) $-1\frac{1}{6}$
5) -2
6) $\frac{3}{8}$
7) $-\frac{7}{12}$
8) $2\frac{1}{2}$
9) $-2\frac{1}{4}$
10) $\frac{1}{8}$
11) 9
12) $\frac{7}{15}$

Review
1) -63
2) 27
3) 24
4) 2

Lesson 1-6
Page 16

Exercises
1) - .91
2) -10.3
3) -12.81
4) 3.17
5) -4.284
6) 12.13
7) 2.37
8) .18
9) .426
10) 5.47
11) 2.01
12) -17.04

Review
1) $\frac{1}{6}$
2) $-\frac{1}{3}$
3) $-1\frac{1}{6}$
4) $-1\frac{1}{8}$

Solutions

Lesson 1-7
Page 17

Exercises
1) 64
2) 16
3) 81
4) 64
5) 625
6) 32
7) 3^5
8) $9^2, 3^4$
9) $8^2, 4^3$
10) $(-2)^3$
11) 11^2
12) $(-1)^3$

Review
1) $1\frac{1}{3}$
2) -4
3) $\frac{1}{10}$
4) 13

Lesson 1-8
Page 18

Exercises
1) 5^{11}
2) $\frac{7^2}{4^2}$
3) $2^4 \times 5^4$
4) 4^6
5) 2^4
6) $\frac{1}{6^2}$
7) 3^5
8) $4^2 \times 5^2$
9) 3^6
10) $\frac{1}{2^4}$
11) $\frac{1}{3^2}$
12) 5^6

Review
1) 25
2) 81
3) -17
4) $-3\frac{3}{4}$

Lesson 1-9
Page 19

Exercises
1) 5
2) 10
3) 30
4) 20
5) $5\sqrt{2}$
6) $2\sqrt{5}$
7) 3
8) $\frac{5}{6}$
9) $\frac{2\sqrt{2}}{5}$
10) 60
11) $3\sqrt{2}$
12) 3

Review
1) -8
2) 6
3) $-\frac{3}{4}$
4) -.4

Lesson 1-10
Page 21

Exercises
1) 28
2) 38
3) 34
4) 7
5) 36
6) 32
7) 29
8) 12
9) 6
10) 9
11) 12
12) 154
13) 9
14) 22
15) 60
16) 4
17) 12
18) 3

Review
1) 9
2) $\frac{4}{5}$
3) 81
4) $3\sqrt{3}$

Lesson 1-11
Page 22

Exercises
1) commutative of +
2) distributive
3) inverse of +
4) associative of x
5) identity of +
6) inverse of x
7) commutative of +
8) commutative of x
9) associative of +
10) identity of x
11) distributive
12) inverse of +

Review
1) $6\frac{1}{2}$
2) 18
3) 38
4) 4

Lesson 1-12
Page 23

Exercises
1) 0
2) -7, -2, 10
3) 5, 9
4) 3, 4
5) 2, 4, 7
6) -5, -8
7) 10, 9, -6, 6
8) 9, -7, 1
9) -8, 8, 0, -3
10) 0, 7, 8
11) 5, 4, -3
12) 5, -3, 9, -8

Review
1) -24
2) -34
3) 42
4) -71

Solutions

Lesson 1-13
Page 25

Exercises

	Part A	Part B
1)	(2, 1)	B
2)	(-4, -2)	A
3)	(6, 3)	C
4)	(2, -5)	D
5)	(-7, -3)	F
6)	(-5, 1)	M
7)	(4, 6)	E
8)	(4, -3)	J
9)	(-3, -5)	H
10)	(-2, 2)	G
11)	(2, 1)	K
12)	(-6, 7)	I

Review
1) -15
2) -4
3) 5
4) -12

Lesson 1-14
Page 27

Exercises
1) no   2 is paired with two y-values
2) yes   each x-value paired with exactly one y-value
3) 2, 6
4) ordered pair
5) If a vertical line can only intersect a graph at 1 point, the graph is a function.
6) 1, 2, 5, 9
7) 2, 3, 6, 7
8) yes   each x-value paired with exactly one y-value
9) 1, 3, 5
10) 2, 4, 7
11) no   3 is paired with two y-values

Review
1) $-\frac{1}{6}$
2) $-\frac{7}{10}$
3) 17
4) -4.2

Lesson 1-15
Page 29

Exercises
1) 2
2) 2, 4, 5, 10
3) 2, 3, 4, 5, 6, 9, 10
4) 2, 3, 4, 6, 9
5) $2 \times 3^2 \times 5$
6) $2^4 \times 3$
7) $2^4 \times 7$
8) $2^2 \times 5^2$
9) $2^2 \times 3 \times 7$
10) $2^2 \times 3 \times 5^2$
11) $2^6 \times 5$
12) $2^5 \times 3$
13) $2^3 \times 3 \times 13$
14) $2^3 \times 3 \times 5$
15) $2^2 \times 5 \times 11$
16) $2 \times 3^2 \times 5^2$

Review
1) -1.7
2) 3^5
3) 2^2
4) 1

Lesson 1-16
Page 30

Exercises
1) 16
2) 4
3) 8
4) 8
5) 10
6) 6
7) 6
8) 14
9) 1

Review
1) 1.7
2) $-\frac{1}{2}$
3) -42
4) $3\sqrt{7}$

Lesson 1-17
Page 31

Exercises
1) 60
2) 180
3) 80
4) 84
5) 24
6) 180
7) 42
8) 360
9) 270

Review
1) -27
2) $3\sqrt{3}$
3) -1.46
4) 16

Solutions

Lesson 1-18
Page 32-33

Exercises
1) 1.236×10^{10}
2) 1.49×10^{-7}
3) 1.597×10^{5}
4) 7.216×10^{-6}
5) 1.096×10^{9}
6) 1.963×10^{-4}
7) 1.6×10^{-9}
8) 7.9×10^{9}
9) 9.3×10^{-5}
10) 9.3×10^{7}
11) 703,200
12) 23,000
13) .00005
14) 11,270
15) 21,000
16) .0021
17) 320,000
18) 7,000,000
19) .00261
20) 356,000

Review
1) -22
2) -36
3) 4
4) 60

Lesson 1-19
Page 35

Exercises
1) $\frac{5}{3}$
2) $\frac{4}{3}$
3) $\frac{6}{5}$
4) $\frac{2}{3}$
5) yes
6) no
7) yes
8) no
9) n = 1
10) n = 2
11) n = 21
12) $n = 2\frac{1}{3}$
13) $n = 2\frac{4}{5}$
14) n = 18
15) $n = 11\frac{2}{3}$
16) $n = \frac{1}{6}$

Review
1) no
2) 0, 2, 7
3) 1, 3, 4, 6
4) 4^2

Lesson 1-20
Page 36-37

Exercises
1) 2 liters
2) $17.50
3) 15 girls
4) 40 kilometers
5) $2\frac{4}{5}$ kilograms
6) $7.14
7) 32.5 km
8) 4.5 cm
9) $23\frac{1}{3}$ hrs = 23 hrs. 20 min.
10) 120 boys
11) 22.5 kg
12) $1500
13) 28 cm
14) $2.00

Review
1) .04
2) -1.6
3) -7.8
4) .71

Lesson 1-21
Page 39

Exercises
1) $.2, \frac{1}{5}$
2) $.09, \frac{9}{100}$
3) $.45, \frac{9}{20}$
4) $.75, \frac{3}{4}$
5) 17.5
6) 4.32
7) 20
8) 180
9) 25%
10) 75%
11) 21.7%
12) 80
13) 20
14) 140
15) 5.6
16) 60%
17) 13.2
18) 24

Review
1) $-\frac{5}{6}$
2) $-1\frac{3}{8}$
3) $\frac{3}{4}$
4) -7

Lesson 1-22
Page 41

Exercises
1) 20
2) 75%
3) $25
4) $1,600
5) 30
6) 90%
7) 250
8) 25
9) $240
10) 180
11) 20%
12) $210

Review
1) $\frac{1}{4}$
2) $-\frac{7}{12}$
3) $-1\frac{1}{2}$
4) 2

Solutions

Chapter 1 Review
Page 42-43

1) -8
2) -9
3) 18
4) -4
5) $-\frac{11}{20}$
6) $-\frac{2}{5}$
7) -2
8) $2\frac{2}{5}$
9) -6.1
10) -2.2
11) -3.91
12) 1.11
13) 81
14) 64
15) $7^2 = 49$
16) 20
17) 43
18) 39
19) 1, 2, 3, 4
20) 2, 7, 8
21) yes
22) $2^2 \times 3^3$
23) 60
24) 120
25) 3.21×10^8
26) 3.62×10^{-6}
27) .000273
28) $x = 2\frac{1}{2}$
29) $n = 9$
30) 13.5
31) 20%
32) 48
33) 36 students
34) $75

Lesson 2-1
Page 45

Exercises
1) $n = 9$
2) $x = 22$
3) $x = -9$
4) $n = 3$
5) $x = 17$
6) $x = 17$
7) $n = -4$
8) $x = -14$
9) $x = 79$
10) $x = -17$
11) $x = -21$
12) $n = 13$
13) $y = 19$
14) $n = 21$
15) $x = 17$
16) $n = -13$
17) $x = -2$
18) $x = -119$

Review
1) 7.2
2) 72
3) 20%
4) 150

Lesson 2-2
Page 47

Exercises
1) $x = 8$
2) $x = 63$
3) $x = -8$
4) $x = -4$
5) $x = 20$
6) $x = -21$
7) $x = 2$
8) $n = 10$
9) $x = 5$
10) $n = -60$
11) $x = 45$
12) $x = -21$
13) $x = -20$
14) $n = 12$
15) $n = 28$
16) $n = -15$
17) $x = 4$
18) $x = -3$

Review
1) $x = 9$
2) $x = 24$
3) $x = 27$
4) $x = 3\frac{3}{5}$

Lesson 2-3
Page 49

Exercises
1) $x = 7$
2) $x = -1$
3) $x = 4$
4) $n = -4$
5) $n = -15$
6) $n = 75$
7) $x = 12$
8) $m = 1$
9) $n = 2$
10) $x = 12$
11) $x = -15$
12) $x = -8$
13) $n = 27$
14) $x = 144$
15) $n = 90$
16) $x = -6$
17) $x = 12$
18) $x = -2$

Review
1) $-\frac{1}{3}$
2) $\frac{7}{15}$
3) -1
4) -5

Lesson 2-4
Page 50-51

Exercises
1) $x = 1$
2) $n = 7$
3) $x = 1$
4) $n = 2$
5) $r = -3$
6) $x = 2$
7) $m = -5$
8) $x = -1$
9) $x = 3$
10) $x = 6$
11) $x = 11$
12) $x = 2$
13) $n = 1$
14) $x = 10$
15) $x = -18$
16) $x = 8$
17) $x = 2$
18) $n = 4$

Review
1) $2 \times 3 \times 5 \times 7$
2) 9.6
3) 9×10^{-6}
4) 21

Solutions

Lesson 2-5
Page 53

Exercises
1) $n = 2$
2) $x = 5$
3) $x = -6$
4) $x = 3$
5) $x = -8$
6) $y = 11$
7) $m = 2$
8) $x = 6$
9) $m = 0$
10) $x = 15$
11) $n = 6$
12) $n = -\frac{3}{2} = -1\frac{1}{2}$
13) $x = -2$
14) $x = \frac{7}{5} = 1\frac{2}{5}$
15) $x = 11$
16) $x = -2$

Review
1) 225
2) 5^2
3) $m = 3$
4) $x = 6$

Lesson 2-6
Page 55

Exercises
1) $x = 6$
2) $x = -3$
3) $a = 3$
4) $x = 8$
5) $x = 29$
6) $n = 1$
7) $x = 5$
8) $n = 1$
9) $x = 0$
10) $x = -10$
11) $x = 4$
12) $y = 4$
13) $x = 7$
14) $y = 19$
15) $x = -\frac{1}{2}$
16) $x = -\frac{1}{8}$

Review
1) 15
2) $x = -9$
3) $x = 7$
4) $x = -7$

Lesson 2-7
Page 57

Exercises
1) $x = 7$ or $x = -7$
2) $x = 6$ or $x = -6$
3) $x = 2$ or $x = -2$
4) $n = 3$ or $n = -3$
5) $x = 2$ or $x = -18$
6) $x = 12$ or $x = -5$
7) $x = 28$ or $x = -4$
8) $x = 5$ or $x = \frac{5}{3} = 1\frac{2}{3}$
9) $x = 6$ or $x = -2$
10) $x = 4$ or $x = -4$
11) $y = \frac{9}{2} = 4\frac{1}{2}$ or $y = -\frac{9}{2} = -4\frac{1}{2}$
12) $x = 4$ or $x = -\frac{8}{3} = -2\frac{2}{3}$
13) $x = 6$ or $x = -7$
14) $x = 0$ or $x = 8$
15) $x = 2$ or $x = 4$
16) $x = 7$ or $x = -\frac{37}{3} = -12\frac{1}{3}$

Review
1) $x = -63$
2) $m = 3$
3) $x = 1$
4) $m = 6$

Lesson 2-8
Page 59

Exercises
1) $3x + 5y$
2) $6x - 3y$
3) $10m + 3x$
4) $-3y$
5) $-x + 5y - 11$
6) $-3m - 12n + 5$
7) $3x + 6y - 3$
8) $-8x - 4xy + 3y + 14$
9) $10x - y$
10) $-5x - 6y$
11) $-7x - 9y$
12) $-2x - 2xy + 8y$
13) $5x + 7y + 5z$
14) $-x^2 - y - x^2y$
15) $4x - 8y + 12$
16) $10x^2 - 11x - xy$

Review
1) $7^1 = 7$
2) n^{15}
3) n^6
4) 11

Lesson 2-9
Page 61

Exercises
1) $x = -11$
2) $n = -14$
3) $x = 3$
4) $n = 0$
5) $x = \frac{3}{2} = 1\frac{1}{2}$
6) $x = -6$
7) $x = 20$
8) $x = 8$
9) $x = -22$
10) $m = 6$
11) $x = -100$
12) $x = 3$
13) $x = 2$
14) $n = 15$

Review
1) $m = 5$
2) $x = 1$
3) $x = 6$
4) $x = 2$

Chapter 2 Review
Page 62

1) $x = 24$
2) $x = -12$
3) $n = 23$
4) $n = -21$
5) $x = 5$
6) $x = -35$
7) $x = 6$
8) $n = 10$
9) $n = 4\frac{1}{2}$
10) $m = 2$
11) $y = -10\frac{1}{2}$
12) $x = 5$
13) $x = 7$ or $x = -13$
14) $x = 4$ or $x = -4$
15) $x = -4$
16) $x = 3$
17) $x = 1$
18) $x = 6$

Solutions

1)
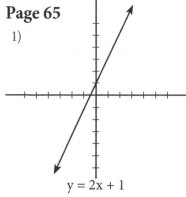
$y = 2x + 1$

2)
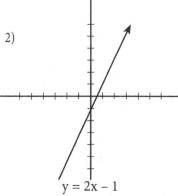
$y = 2x - 1$

3)
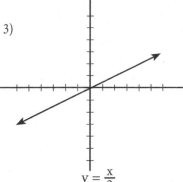
$y = \frac{x}{2}$

4)
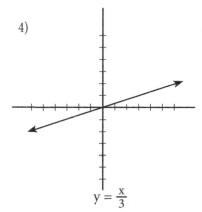
$y = \frac{x}{3}$

5)
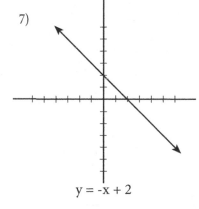
$y = \frac{x}{2} + 5$

6)
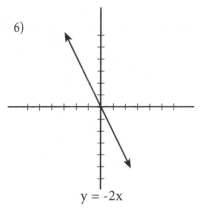
$y = -2x$

7)
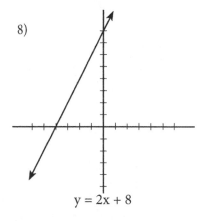
$y = -x + 2$

8)
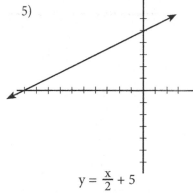
$y = 2x + 8$

9)
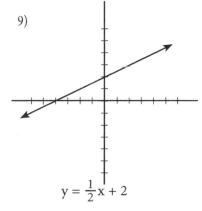
$y = \frac{1}{2}x + 2$

10)
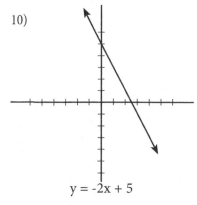
$y = -2x + 5$

11)
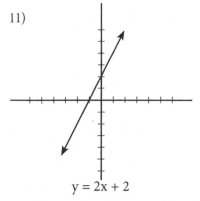
$y = 2x + 2$

12)
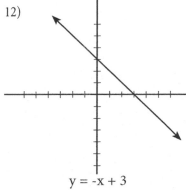
$y = -x + 3$

Solutions

Lesson 3-1, continued

13)

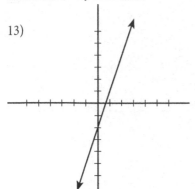

$y = 3x - 2$

14)

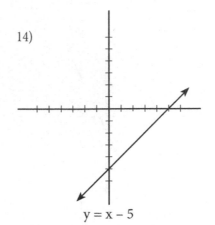

$y = x - 5$

Review
1) $x = 6$ or $x = -14$
2) $x = 14$ or $x = -6$
3) $x = 5$ or $x = -5$
4) $x = 7$ or $x = -7$

Lesson 3-2
Page 67

1)

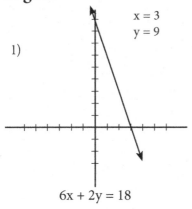

$x = 3$
$y = 9$

$6x + 2y = 18$

2)

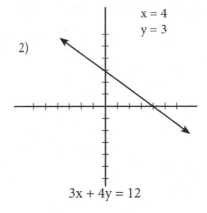

$x = 4$
$y = 3$

$3x + 4y = 12$

3)

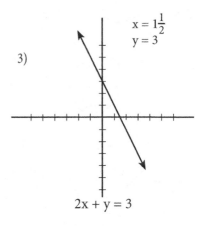

$x = 1\frac{1}{2}$
$y = 3$

$2x + y = 3$

4)

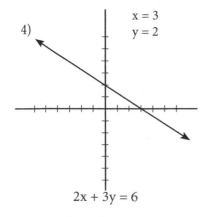

$x = 3$
$y = 2$

$2x + 3y = 6$

5)

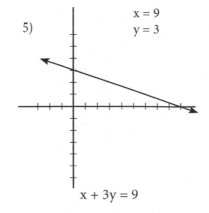

$x = 9$
$y = 3$

$x + 3y = 9$

6)

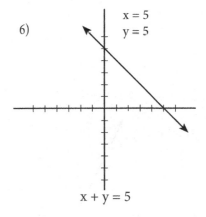

$x = 5$
$y = 5$

$x + y = 5$

Solutions

7) 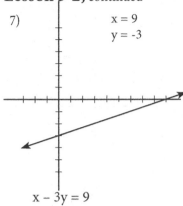 x = 9
y = -3

x – 3y = 9

8) 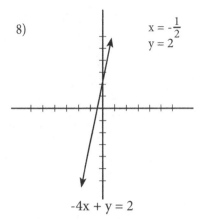 $x = -\frac{1}{2}$
y = 2

-4x + y = 2

9) 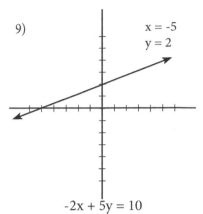 x = -5
y = 2

-2x + 5y = 10

10) 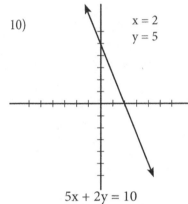 x = 2
y = 5

5x + 2y = 10

11) 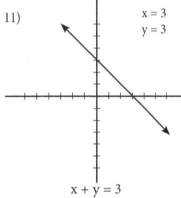 x = 3
y = 3

x + y = 3

12) 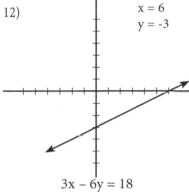 x = 6
y = -3

3x – 6y = 18

13) 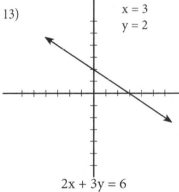 x = 3
y = 2

2x + 3y = 6

14) 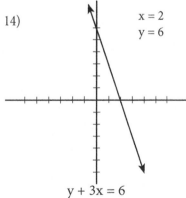 x = 2
y = 6

y + 3x = 6

15) 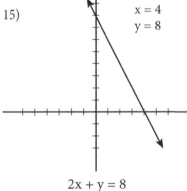 x = 4
y = 8

2x + y = 8

16) 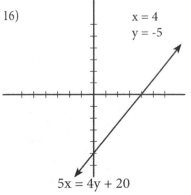 x = 4
y = -5

5x = 4y + 20

Review
1) 25
2) 80
3) 8
4) 25%

Solutions

Lesson 3-3
Page 69

Exercises

1) $\frac{1}{3}$ 14) $\frac{5}{4}$

2) $\frac{3}{2}$ 15) $\frac{1}{3}$

3) $-\frac{3}{2}$ 16) $\frac{7}{10}$

4) $\frac{1}{3}$ 17) $\frac{2}{7}$

5) -2   18) $\frac{1}{8}$

6) $\frac{5}{8}$

7) $\frac{1}{2}$

8) $\frac{7}{2}$

9) $\frac{6}{5}$

10) $\frac{4}{5}$

11) $\frac{3}{2}$

12) $-\frac{3}{2}$

13) $\frac{1}{2}$

Review

1)

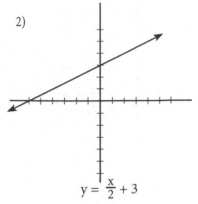

$$y = 2x + 4$$

2)

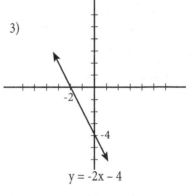

$$y = \frac{x}{2} + 3$$

3)

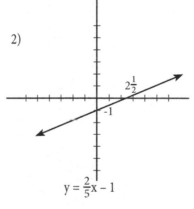

$$y = \frac{x}{3} + 1$$

4)

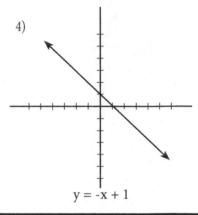

$$y = -x + 1$$

Lesson 3-4
Page 71

1)

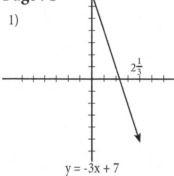

$$y = -3x + 7$$

2)

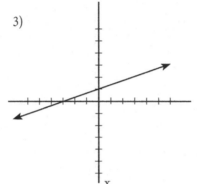

$$y = \frac{2}{5}x - 1$$

3)

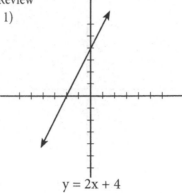

$$y = -2x - 4$$

4)

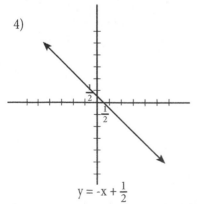

$$y = -x + \frac{1}{2}$$

5)

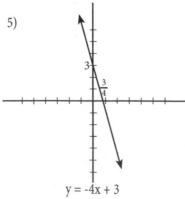

$$y = -4x + 3$$

6)

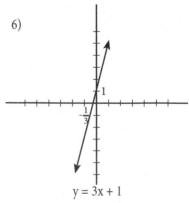

$$y = 3x + 1$$

Solutions

Lesson 3-4, continued

7)

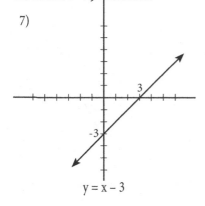

$$y = x - 3$$

8)

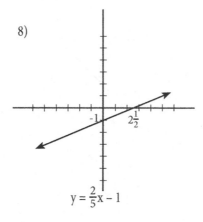

$$y = \frac{2}{5}x - 1$$

9)

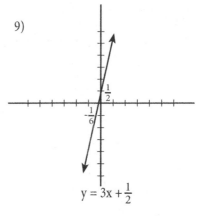

$$y = 3x + \frac{1}{2}$$

10)

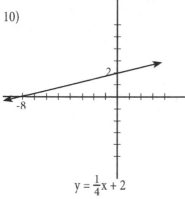

$$y = \frac{1}{4}x + 2$$

11)

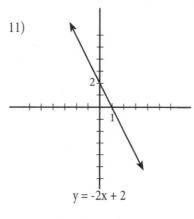

$$y = -2x + 2$$

12)

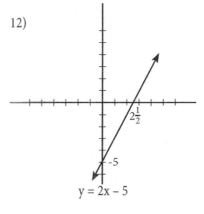

$$y = 2x - 5$$

13)

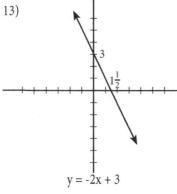

$$y = -2x + 3$$

14)

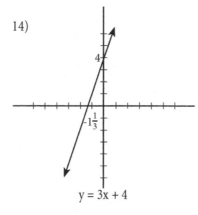

$$y = 3x + 4$$

15)

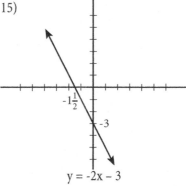

$$y = -2x - 3$$

16)

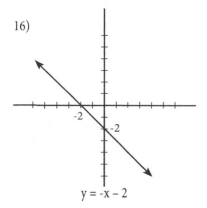

$$y = -x - 2$$

Review
1) $x = -1, y = 1$
2) $x = 2, y = -2$
3) $x = 0, y = 0$
4) $x = \frac{1}{2}, y = -1$

Solutions

Lesson 3-5
Page 73

1)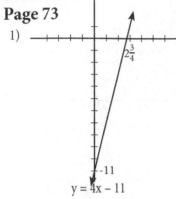

$2\frac{3}{4}$

-11

$y = 4x - 11$

2)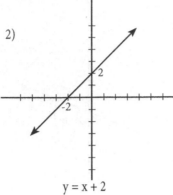

2

-2

$y = x + 2$

3)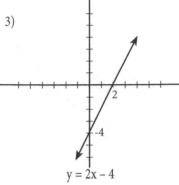

2

-4

$y = 2x - 4$

4)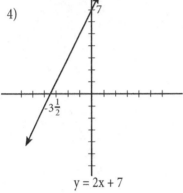

7

$-3\frac{1}{2}$

$y = 2x + 7$

5)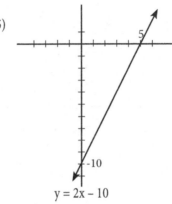

5

-10

$y = 2x - 10$

6)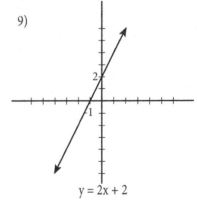

4

-8

$y = \frac{1}{2}x + 4$

7)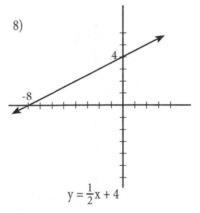

1

-1

$y = x - 1$

8)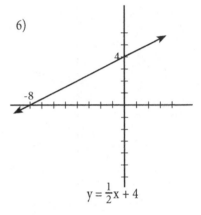

4

-8

$y = \frac{1}{2}x + 4$

9)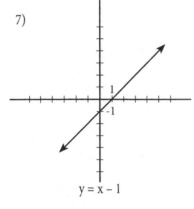

2

-1

$y = 2x + 2$

10)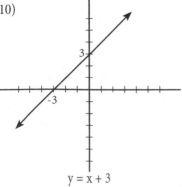

3

-3

$y = x + 3$

11)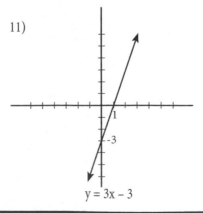

1

-3

$y = 3x - 3$

12)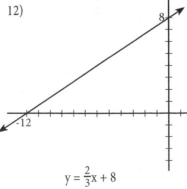

8

-12

$y = \frac{2}{3}x + 8$

Solutions

Lesson 3-5, continued

13)

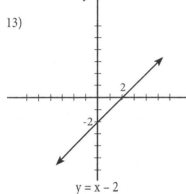

$y = x - 2$

14)

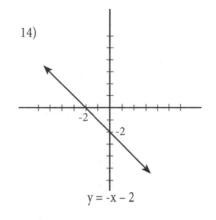

$y = -x - 2$

Review
1) 24
2) 20%
3) 60
4) 20%

Lesson 3-6
Page 75

Exercises
1) $y - 3 = 5(x - 1)$
2) $y - 2 = 3(x - 1)$
3) $y + 3 = 4(x - 2)$
4) $y - 4 = 3(x - 2)$
5) $y - 5 = \frac{3}{4}(x + 1)$
6) $y - 8 = \frac{3}{2}(x - 6)$ or $y - 5 = \frac{3}{2}(x - 4)$
7) $y + 4 = -1(x - 5)$ or $y - 2 = -1(x + 1)$
8) $y - 4 = 3(x - 9)$ or $y - 1 = 3(x - 8)$
9) $y - 3 = -\frac{1}{4}(x + 2)$ or $y - 4 = -\frac{1}{4}(x + 6)$
10) $y - 4 = \frac{3}{2}(x - 7)$ or $y - 7 = \frac{3}{2}(x - 9)$
11) $y - 2 = 5(x - 1)$
12) $y - 4 = -\frac{1}{2}(x - 3)$
13) $y - 2 = x - 3$ or $y + 5 = x + 4$
14) $y + 7 = (x + 8)$ or $y + 3 = (x + 4)$

Review
1) $x = 6$
2) $x = -8$
3) $x = 4$
4) $x = 1$

Chapter 3 Review
Page 76

1)

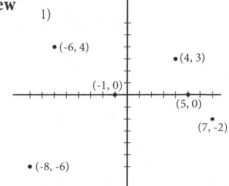

- (-6, 4)
- (4, 3)
- (-1, 0)
- (5, 0)
- (7, -2)
- (-8, -6)

2)

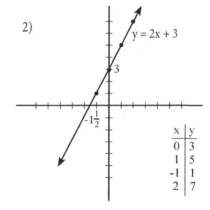

$y = 2x + 3$

3

$-1\frac{1}{2}$

x	y
0	3
1	5
-1	1
2	7

3)

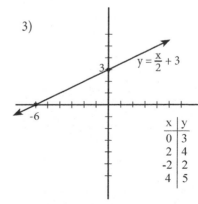

3

$y = \frac{x}{2} + 3$

-6

x	y
0	3
2	4
-2	2
4	5

Solutions

Chapter 3 Review, continued

4) x-intercept = 3
   y-intercept = 2

5) x-intercept = 4
   y-intercept = 8

6) -2

7) $\frac{1}{5}$

8)

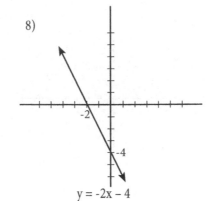

$y = -2x - 4$

9)

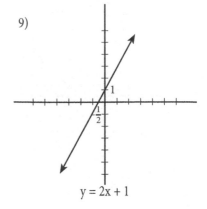

$y = 2x + 1$

10)

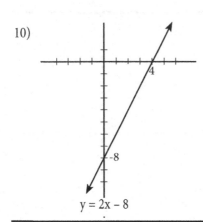

$y = 2x - 8$

11)

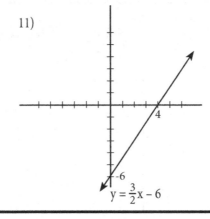

$y = \frac{3}{2}x - 6$

12) $y + 5 = 3(x - 1)$

13) $y - 2 = 2x$ or $y = 2(x + 1)$

Lesson 4-1
Page 79

Exercises

1)

2)

3)

4)

5)

6)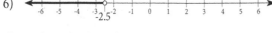

7)

8)

9)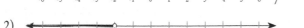

10)

11)

12)

13)

14)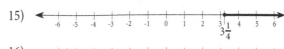

15)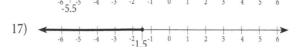

16)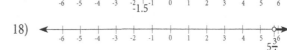

17)

18)

Solutions

Lesson 4-1, continued

Review

1)

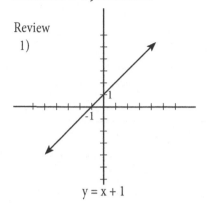

$y = x + 1$

2)

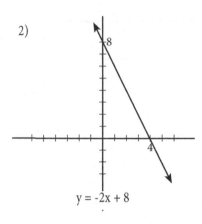

$y = -2x + 8$

3)

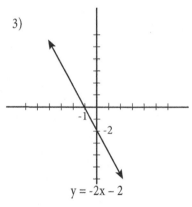

$y = -2x - 2$

4)

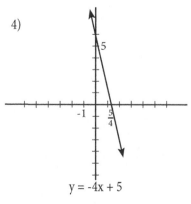

$y = -4x + 5$

Lesson 4-2
Page 81

Exercises

1) $x < 3$

2) $x > -1$

3) $x \geq 2$

4) $x \leq -1$

5) $x < -2$

6) $x > -5$

7) $x \leq -2$

8) $x \geq 2$

9) $x \leq 6$

10) $x > -12$

11) $x < 5$

12) $x < 3\frac{1}{2}$

13) $x < -1$

14) $x \geq -13$

15) $x > 3$

16) $x \geq 1$

17) $x > 2$

18) $x \leq -24$

Review

1) $y = 5x - 20$
2) $y = x$

Solutions

Lesson 4-3
Page 83

Exercises

1) $x \geq -5$ [number line, closed circle at -5, shaded right]
2) $x \geq 4$ [number line, closed circle at 4, shaded right]
3) $x \geq 40$ [number line, closed circle at 40, shaded right]
4) $x > -28$ [number line, open circle at -28, shaded right]
5) $x > -4$ [number line, open circle at -4, shaded right]
6) $x > -4$ [number line, open circle at -4, shaded right]
7) $x > -8$ [number line, open circle at -8, shaded right]
8) $x > -3$ [number line, open circle at -3, shaded right]
9) $n \leq -15$ [number line, closed circle at -15, shaded left]
10) $n \geq 4$ [number line, closed circle at 4, shaded right]
11) $x < 2$ [number line, open circle at 2, shaded left]
12) $x \leq 16$ [number line, closed circle at 16, shaded left]
13) $x > 5$ [number line, open circle at 5, shaded right]
14) $y \geq 3$ [number line, closed circle at 3, shaded right]
15) $x \geq 4$ [number line, closed circle at 4, shaded right]
16) $x > -2$ [number line, open circle at -2, shaded right]
17) $x \leq 8$ [number line, closed circle at 8, shaded left]
18) $x \leq -27$ [number line, closed circle at -27, shaded left]

Review

1) 7^6
2) 7^6
3) 7^{24}
4) n^{12}

Lesson 4-4
Page 85

Exercises

1) $x > -2$ [number line, open circle at -2, shaded left]
2) $x < 3$ [number line, open circle at 3, shaded left]
3) $x \geq -4\frac{1}{2}$ [number line, closed circle at -4½, shaded right]
4) $x > 3$ [number line, open circle at 3, shaded left]
5) $m \geq -21$ [number line, closed circle at -21, shaded right]
6) $m \leq -5$ [number line, closed circle at -5, shaded left]
7) $x \geq 3$ [number line, closed circle at 3, shaded right]
8) $-2 < x = x > -2$ [number line, open circle at -2, shaded right]
9) $x \leq 1$ [number line, closed circle at 1, shaded left]
10) $x > -12$ [number line, open circle at -12, shaded left]

Review

1) $-\frac{3}{5}$
2) -1.6
3) -17
4) $.5$

Solutions

Lesson 4-5
Page 87

Exercises

1) $-3 < x < 2$

2) $-12 \le x < -3$

3) $x > 1$ or $x < -7$

4) $-3 \le x \le 2$

5) $-2 \le x \le 1$

6) $x \le -7$ or $x \ge -3$

7) $-5 < x < 4$

8) $x > 2$ or $x < -5$

9) $-3 < x \le 4$

10) $x \le -2$ or $x > 1$

Review

1) x^8

2) $27n^8$

3) x^{12}

4) 6^7

Lesson 4-6
Page 89

Exercises

1) $x < -10$ or $x > 10$

2) $-10 \le x \le 10$

3) $-6 < x < 6$

4) $x > 2$ or $x < -10$

5) $-4 < x < 16$

6) $x \ge 6$ or $x \le 0$

7) $0 \le x \le 12$

8) $x < -3$ or $x > 3$

9) $-4 < x < 4$

10) $x \le -3$ or $x \ge 3$

11) $7 < x < 15$

12) $-3 \le n \le 8$

13) $y \le -16$ or $y \ge -2$

14) $\frac{3}{2} < x < 3$

15) $-15 < n < 21$

16) $x \le -\frac{2}{3}$ or $x \ge 2\frac{2}{3}$

17) $0 \le n \le 6$

18) $-2 < n < 1$

Review

1) $-\frac{1}{4}$

2) $-\frac{3}{4}$

3) $-\frac{3}{4}$

4) -7

Solutions

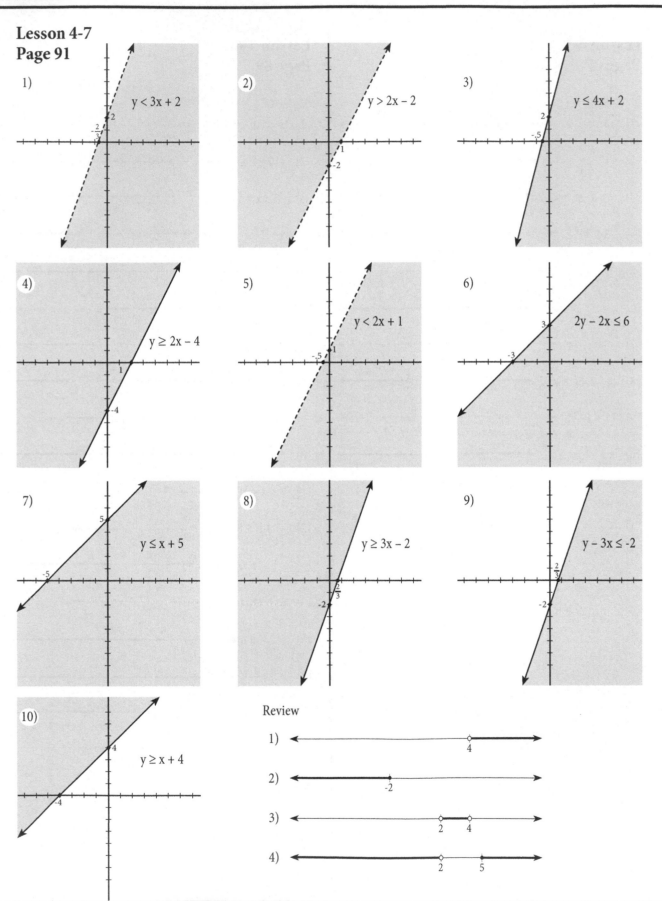

1) $y < 3x + 2$

2) $y > 2x - 2$

3) $y \leq 4x + 2$

4) $y \geq 2x - 4$

5) $y < 2x + 1$

6) $2y - 2x \leq 6$

7) $y \leq x + 5$

8) $y \geq 3x - 2$

9) $y - 3x \leq -2$

10) $y \geq x + 4$

Review

1)

2)

3)

4)

Solutions

Solutions

Chapter 4 Review
Page 92

1)
4

2) ← ——————————
-3

3) ←—○——————————
-5

4) ←——————○———————
1

5) x > -1   ←————○————————→
-1

6) x ≤ 5   ←————————•————→
5

7) x ≥ -4   ←————•————————→
-4

8) x > -5   ←——○————————→
-5

9) x ≥ -4$\frac{1}{2}$ ←——•————————→
-4$\frac{1}{2}$

10) x ≥ -84   ←•————————→
-84

11) x ≤ -13 or x ≥ 3   ←——•————•———→
-13   3

12) x < -8 or x > 8   ←——○————————○——→
-8   8

13) -8 ≤ x ≤ 8   ←——•————————•——→
-8   8

14) x < -6 or x > 2   ←——○————————○——→
-6   2

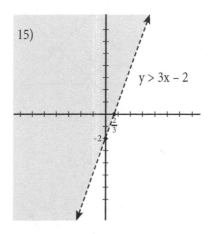

15)

$y > 3x - 2$

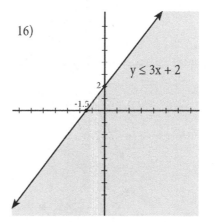

16)

$y ≤ 3x + 2$

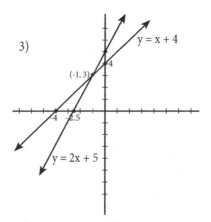

Lesson 5-1
Page 95

1)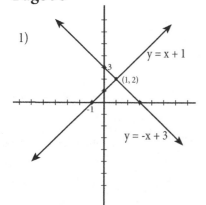

$y = x + 1$
(1, 2)
3
-1
$y = -x + 3$

2)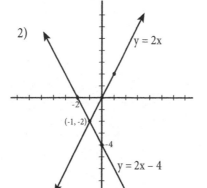

$y = 2x$
-2
(-1, -2)
-4
$y = 2x - 4$

3)

$y = x + 4$
4
(-1, 3)
-4   -2.5
$y = 2x + 5$

Solutions

Lesson 5-1, continued

4)
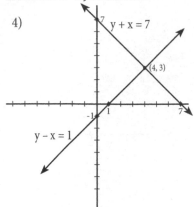
$y + x = 7$
$(4, 3)$
$y - x = 1$

5)
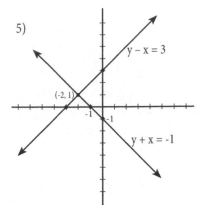
$y - x = 3$
$(-2, 1)$
$y + x = -1$

6)
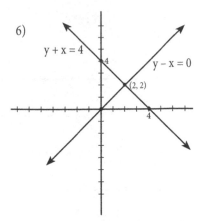
$y + x = 4$
$y - x = 0$
$(2, 2)$

7)
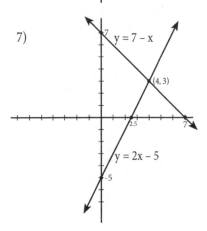
$y = 7 - x$
$(4, 3)$
$y = 2x - 5$

8)
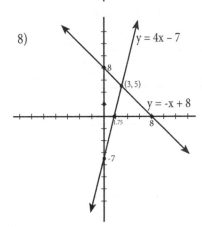
$y = 4x - 7$
$(3, 5)$
$y = -x + 8$

9)
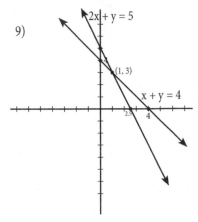
$2x + y = 5$
$(1, 3)$
$x + y = 4$

10)
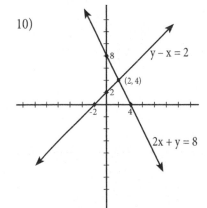
$y - x = 2$
$(2, 4)$
$2x + y = 8$

11)
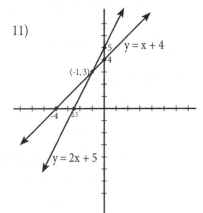
$y = x + 4$
$(-1, 3)$
$y = 2x + 5$

12)
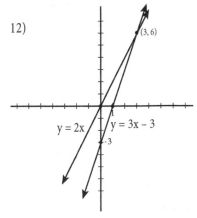
$(3, 6)$
$y = 2x$
$y = 3x - 3$

13)
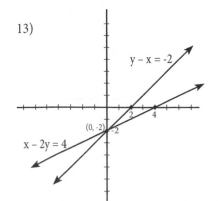
$y - x = -2$
$(0, -2)$
$x - 2y = 4$

14)
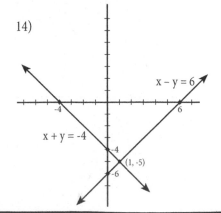
$x - y = 6$
$x + y = -4$
$(1, -5)$

Review

1) $x > 8$

2) $x \geq -4\frac{1}{2}$

3) $x \geq 14$

4) $x > 9$

Solutions

Lesson 5-2
Page 97

Exercises

1) (13, 10)
2) (2, 4)
3) (-1, -3)
4) (3, 7)
5) (3, -1)
6) (1, 4)
7) (1, 4)
8) (-4, -2)
9) (6, 2)
10) $(-10, 3\frac{1}{3})$
11) (4, 0)
12) (2, 7)

Review

1) $x < -10$ or $x > 6$

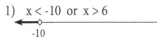

2) $-3 \le x \le 9$

3) $-8 < x < 8$

4) $x < -3$ or $x > 3$

Lesson 5-3
Page 99

Exercises

1) (-2, 2)
2) (3, -2)
3) (3, 4)
4) (-2, -3)
5) (4, -2)
6) (1, 3)
7) (4, 7)
8) (2, 5)
9) (-4, -2)
10) (2, 0)
11) (2, 1)
12) (1, 3)
13) (5, 2)
14) (1, 4)
15) (4, 5)

Review

1)

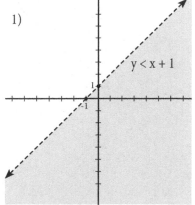

$y < x + 1$

2)

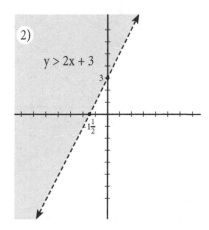

$y > 2x + 3$

Lesson 5-4
Page 101

1)

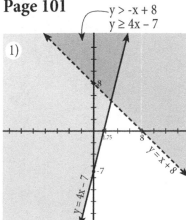

$y > -x + 8$
$y \ge 4x - 7$

2)

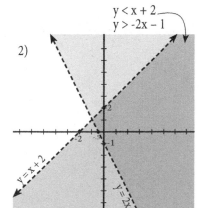

$y < x + 2$
$y > -2x - 1$

3)

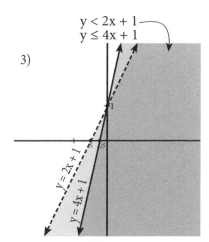

$y < 2x + 1$
$y \le 4x + 1$

Solutions

Lesson 5-4, continued

4)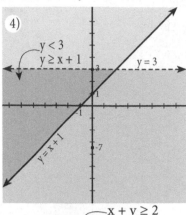
y < 3
y ≥ x + 1
y = 3
y = x + 1

5)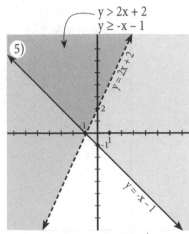
y > 2x + 2
y ≥ -x – 1
y = 2x + 2
y = -x – 1

6)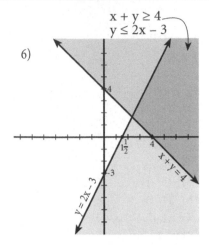
x + y ≥ 4
y ≤ 2x – 3
x + y = 4
y = 2x – 3

7)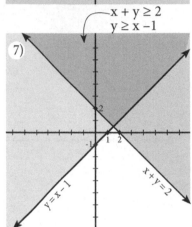
x + y ≥ 2
y ≥ x –1
y = x – 1
x + y = 2

8)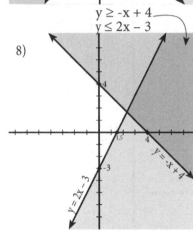
y ≥ -x + 4
y ≤ 2x – 3
y = 2x – 3
y = -x + 4

Review
1) (2, 8)
2) (7, 5)

Chapter 5 Review
Page 102

1) (1, 2)
2) (-3, 0)
3) (3, 6)
4) (2, 3)
5) (1, 4)
6) (3, -1)
7) (1, -4)
8) (-1, 3)
9) (2, 0)
10) (3, 1)

11)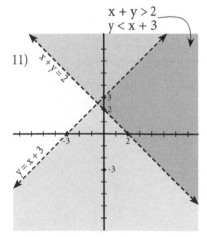
x + y > 2
y < x + 3
x + y = 2
y = x + 3

12)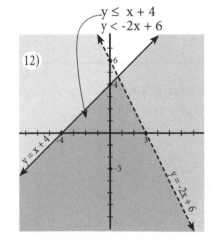
y ≤ x + 4
y < -2x + 6
y = x + 4
y = -2x + 6

Solutions

Lesson 6-1
Page 105

Exercises
1) $22x^2 - 10xy + 12x$
2) $6xy + 2x - 2y$
3) $6x - 9$
4) $-2y + 6$
5) $-11x^2 + 13x + 11$
6) $-y^2 + y - 2x - 5$
7) $11x^2 - 15x - 16$
8) $3x^3 + 5x^2 + 10$
9) $x^3 - 3x^2 - 5x$
10) $x^2y - 5xy - 2x + 11y$
11) $-2x^2 + 2x - 8$
12) $7x^3 + 2x^2 - 6x + 7$
13) $-4x^2 - 3x + 4$
14) $11x^2 + 7x - 24$

Review
1) $(9, 27)$
2) $(3, 8)$

Lesson 6-2
Page 107

Exercises
1) $-2x + 27$
2) $6x^2 + x + 15$
3) $-2x^2 + 4xy + 2z$
4) $4n$
5) $3x^2 + 3$
6) $-2x + y$
7) $8x - 10y$
8) $2x^2 - 7x + 16$
9) $-2x - 12$
10) $3x^2 + 2x + 10$
11) $x - 1$
12) $2x^2 + 9x - 3y^2$
13) $-4x^2 - 7x - 1$
14) $3x - 75$
15) $7x^2 - 22x - 13$
16) $-x^2 - x + 9$
17) $4x^2 + 7$
18) $-x^2 - 11x - 11$
19) $7x^2 - 2x - 19$
20) $-2x^2 - 4xy + 9$

Review
1) 1
2) $-\dfrac{1}{5}$
3) $\dfrac{5}{2}$
4) 2

Lesson 6-3
Page 109

Exercises
1) a^7
2) $3x^7$
3) $-6x^5$
4) $3xy^4$
5) $-24x^6y^5$
6) $12r^{10}s^{10}$
7) $8a^6$
8) $16a^2$
9) $-5m^6n^3p^5$
10) $27a^6$
11) $12x^5y^3$
12) $36x^2y^2$
13) $81x^2$
14) $-40x^2y^4$
15) $-21x^6y^7$
16) $24y^9$
17) $-8x^6y^3$
18) $36x^2y^4$
19) $6x^{10}$
20) $9x^8$

Review
1) $5x^4 - x^3 - x^2 + 2x - 2$
2) $-2m^4 - 4m^2 + 5m - 5$

Lesson 6-4
Page 111

Exercises
1) $6a$
2) $4a^2$
3) $3xy$
4) $\dfrac{4y^2}{x^2}$
5) x^2y^4
6) $-\dfrac{4a}{b}$
7) $6x$
8) $\dfrac{9a}{c^2}$
9) $\dfrac{x^2}{yz^3}$
10) $5x^5y^5z$
11) $\dfrac{4x}{y^2}$
12) $-3xyz^3$
13) $\dfrac{-5y^2}{x^2z^2}$
14) $\dfrac{9}{xy}$
15) $\dfrac{y}{x^2}$
16) $3x^8y^4$
17) $\dfrac{3m}{n^2}$
18) $15m^2n$
19) $\dfrac{5y^2}{z}$
20) $3x^5y^2$

Review
1) $5x^4 + 3x^3 - 3x^2 + 12x - 8$
2) $-2y^5 + 5y^3 + 14y + 38$

Lesson 6-5
Page 113

Exercises
1) $8bc^2 + 20abc$
2) $45ac^2 + 10c^2$
3) $50m^6n - 50m^4n^4$
4) $15x^3yz - 5x^2y^2z + 5xyz^3$
5) $-10x^3 + 35x^2 - 35x$
6) $-32x - 28y$
7) $6x - 15x^2 + 15x^3$
8) $21m^3 - 49m^2 + 21m$
9) $-12x^3 + 21x^2 - 9x$
10) $-12x^4y^2 + 6x^3y^3 - 6x^3y^2$
11) $12x^4 - 8x^3 + 4x$
12) $-18x^4y + 6x^2y^2 + 12xy^2$
13) $4y^3z - 12y^2z^2 + 4yz^3$
14) $-24a^3b^4 + 9a^2b^4$
15) $-6r^4s^2 + 12r^3s^3 - 9r^2s^4$
16) $12x^3y - 30x^2y^2$
17) $5x^3 - 7x^4 + 3x^5 - x^7$
18) $-60x^2 + 20xy - 80$
19) $-3x^3y^2 + 12x^2y - 3x^4$
20) $2x^4y - 6x^2y^2 + 2xy^3$

Review
1) x^3y^2
2) $-6x^6$
3) $20x^3y^2$
4) $-6a^3b^6c^3$

Lesson 6-6
Page 115

Exercises
1) $x^2 + 8x + 15$
2) $x^2 - x - 12$
3) $x^2 - 8x + 15$
4) $3x^2 + 17x + 10$
5) $2x^2 - 13x + 6$
6) $14x^2 - x - 3$
7) $10x^2 - 11x + 3$
8) $6x^2 + 13x + 6$
9) $25x^2 + 40x + 16$
10) $2x^2 - x - 6$
11) $-4x^2 - 2x + 2$
12) $3x^2 + 14x + 8$
13) $6x^2 + 17xy + 12y^2$
14) $10x^2 - 11xy + 3y^2$
15) $12m^2 - 12mn + 3n^2$
16) $10x^2 + 11xy - 6y^2$

Review
1) $4x^2$
2) $-4x^2$
3) $-3a^4b^4$
4) -6

Solutions

Lesson 6-7
Page 117

Exercises
1) 4
2) $2x + 3 - 3y$
3) $y^2 + yz + z^2$
4) $2x^2 + 3x - 1$
5) $2xy + 3y^2 - 6$
6) $3y + 2 - 3x$
7) $2xy - 3x^3 + 5y^2$
8) $7x^3y^2 - 3x^4y^3 + 7x^2y$
9) $6x^4y - 4x^2 + 3y$
10) $4x^3y^3 + 2x^2y^2 + 2xy + 2$
11) $-2x - 3y + 5$
12) $2x - 3y + x^2$
13) $-x^2 + x - 2$
14) $3y - 2 + 5x$
15) $3x - 6x^2 + 2$
16) $3xy - 5 + 6x$

Review
1) $-12x^7$ 3) $3a^5b^5c$
2) xy 4) $-7x^3y$

Lesson 6-8
Page 119

Exercises
1) $x + 2$ 13) $x + 1$
2) $x + 4$ 14) $n - 1$
3) $x + 2 + \dfrac{-2}{x - 6}$ 15) $3x + 2 + \dfrac{2}{x + 4}$
4) $x - 4$ 16) $5x - 1$
5) $x - 6 + \dfrac{2}{x - 3}$
6) $x - 7 + \dfrac{-7}{x - 2}$
7) $x - 7 + \dfrac{2}{x - 1}$
8) $x - 18 + \dfrac{-5}{x + 3}$
9) $x - 2$
10) $7x - 1$
11) $2x - 5 + \dfrac{2}{3x + 2}$
12) $x - 1$

Review
1) $4x^4 + 11x^3 + 17x^2 + 17x + 4$
2) $6m^4 + 3m^3 + 2m^2 + 8m - 5$

Lesson 6-9
Page 121

Exercises
1) $7x(4x - 1)$
2) $4x(x - 2)$
3) $12x^2(1 + 4y)$
4) $7n(3n - 2m)$
5) $6(x^2 + 2x + 4xy + 6)$
6) $8ab^3(4ab - 2 + 5a^2b^2)$
7) $7mn(3 + 4mn + 5m^2n^2)$
8) $4ab^2(ab^5 - 8ab^4 + 2)$
9) $7xy(3x^2y - x + 6)$
10) $5a(a^3 + 5a^2 - 7a + 4)$
11) $3y(x^4 - 6y^2)$
12) $4x^3y^3(x^2 + 1)$
13) $3xy(y - 6)(y + 1)$
14) $x(x^2 + x + 2)$
15) $2(x^2 + 4x + 2)$
16) $x(x^2 - 4x - 4)$
17) $3x(5x^2 - 3x + 1)$
18) $x(16x^2 + 25y^2)$
19) $5x(x + 10x^2 - 25)$
20) $3x(x^2 + 10x - 25)$

Review
1) $4m^2 + 8mn + 12m$
2) $-15x^2 - 10xy$
3) $-x^3 - 2x^4$
4) $-4c^3 - 8c^2 + 10c$

Lesson 6-10
Page 123

Exercises
1) $(x + 1)(x + 3)$
2) $(x + 3)(x + 8)$
3) $(x + 4)(x + 8)$
4) $(x + 3)(x + 21)$
5) $(x - 2)(x - 3)$
6) $(x - 9)(x - 3)$
7) $(x - 4)(x + 2)$
8) $(x + 8)(x - 3)$
9) $(x - 8)(x + 6)$
10) $(x + 2)(x + 13)$
11) $(x - 8)(x + 4)$
12) $(x - 5)(x - 8)$
13) $(x + 5)(x - 1)$
14) $(x + 1)(x + 1)$
15) $(x - 4)(x + 2)$
16) $(x - 4)(x - 12)$
17) $(x + 1)(x - 13)$
18) $(x + 8)(x - 5)$

Review
1) $x^2 + 5x + 6$
2) $2x^2 + 5xy + 2y^2$
3) $y^2 - y - 20$
4) $6y^2 - y - 2$

Lesson 6-11
Page 125

Exercises
1) $(x + 11)(x - 11)$
2) $(2x + 5y)(2x - 5y)$
3) $(2y + 1)(2y - 1)$
4) $(10y + a)(10y - a)$
5) $(4y + 9)(4y - 9)$
6) $(x + y)(x - y)$
7) $(12 + x)(12 - x)$
8) $(8x + 2y)(8x - 2y)$
9) $(10x + 9y)(10x - 9y)$
10) $(1 + x)(1 - x)$
11) $(4x + 5y)(4x - 5y)$
12) $(11 + m)(11 - m)$
13) $(5x^2 + y^2)(5x^2 - y^2)$
14) $(mn + 11)(mn - 11)$
15) $(5m^3 + 4y^3)(5m^3 - 4y^3)$
16) $(6ab + xy)(6ab - xy)$
17) $(4mn + 3xy)(4mn - 3xy)$
18) $(6x^3 + 5y^4)(6x^3 - 5y^4)$

Review
1) $a + 3b$
2) $4x - 22$
3) $x^3 + x^2 - 2x$
4) $a + 9$

Solutions

Lesson 6-12
Page 127

Exercises
1) $x(y + z)(y - z)$
2) $2(x + 4)(x - 6)$
3) $x(x + 2)(x + 5)$
4) $2(x + 10y)(x - 10y)$
5) $2(3x + 2)(3x - 2)$
6) $3(x + 1)(x + 9)$
7) $2a(x + 2)(x - 3)$
8) $4(a + 3)(a - 3)$
9) $3(x + 4)(x - 4)$
10) $4(x - 5)^2$
11) $2(2x + 1)(x - 2)$
12) $y(y + 5)(y - 5)$
13) $x(x + 2)(x + 5)$
14) $2(x + 4)(x - 4)$
15) $4(x + 3)(x - 4)$
16) $m(x + y)(x - y)$
17) $2(x + 4)(x - 1)$
18) $a^2(2y - 1)(y - 1)$

Review
1) $x + 2$
2) $2x + 5y^2 + 2xy^2$

Lesson 6-13
Page 129

Exercises
1) $(2x + 1)(x + 2)$
2) $(3x + 4)(x + 2)$
3) $(2x - 1)(x - 1)$
4) $(3x - 2)(x - 2)$
5) $(3x + 4)(x - 3)$
6) $(5x - 8)(x + 1)$
7) $(2x - 3)(x + 2)$
8) $(2x - 1)(x + 3)$
9) $(3x - 2)(x + 1)$
10) $(2x + 3)(x - 3)$
11) $(2x + 1)(x + 5)$
12) $(2x + 3)(x + 4)$
13) $(5x - 1)(x - 3)$
14) $(3x - 7)(x + 2)$
15) $(3x - 5)(x - 2)$
16) $(7x - 2)(x - 1)$
17) $(2x + 5)(x + 3)$
18) $(2x + 4)(2x - 3)$

Review
1) $x^2 + 5x + 6$
2) $x^2 + 3x - 28$
3) $2x^2 - 7x - 4$
4) $6x^2 - 5x - 6$

Chapter 6 Review
Page 130

1) $25x^2 + 2xy + 12x$
2) $7x^3 + 12x^2 + 2x + 8$
3) $35x^5y^3$
4) $12x^5y^6$
5) $\frac{5a^4}{b^2}$
6) $x^4y + 3x^2y^2 - xy$
7) $6x^2 - 5x - 21$
8) $7x + 1$
9) $8x(3 - x)$
10) $(x + 5)(x + 5) = (x + 5)^2$
11) $2(x + 2)(x + 4)$
12) $(x + 7)(x - 7)$
13) $(7x + 9y)(7x - 9y)$
14) $2(7x + 10y)(7x - 10y)$
15) $xy(x + y)(x - y)$
16) $a^2(b + 6)^2$ or $a^2(b + 6)(b + 6)$
17) $(3x + 2)(x - 4)$
18) $(3x + 1)(2x - 3)$

Lesson 7-1
Page 133

Exercises
1) $2m$
2) $\frac{b}{3a}$
3) $-\frac{4}{xy}$
4) $\frac{3n}{r}$
5) $-\frac{5x}{3y^2z^6}$
6) $\frac{9m}{8}$
7) $\frac{4b^3}{a}$
8) $\frac{3}{xyz}$
9) $\frac{9}{xy^2}$
10) $\frac{s}{r^2t}$
11) $15x$
12) $\frac{1}{5x^2y^3}$
13) $\frac{8x}{3}$
14) $-\frac{6c}{ab}$
15) $-\frac{5y}{7x}$
16) $\frac{1}{4x^2y^2}$
17) $\frac{3r}{m^2n}$
18) $\frac{8}{x^3y^3z^3}$

Review
1) $2(x + 2)(x + 5)$
2) $4(x - 5)^2$
3) $8(x + 2)(x - 2)$
4) $2(x + 9)(x - 9)$

Lesson 7-2
Page 135

Exercises
1) $\frac{1}{x}$
2) $\frac{1}{3x - 2}$
3) $x - y$
4) $2m$
5) $\frac{x - 4}{3}$
6) $\frac{6}{x - y}$
7) $x - 5$
8) $\frac{x + 2}{x + 3}$
9) $\frac{x - 2}{x + 2}$
10) $\frac{x + 1}{x + 3}$
11) $\frac{x - 3}{x + 3}$
12) $\frac{x + 5}{x - 3}$
13) $\frac{x + y}{x - y}$
14) $\frac{3}{y + 2}$
15) $\frac{x + 2}{x - 3}$
16) $\frac{2x - 1}{x - 3}$
17) $\frac{x + y}{x - y}$
18) $\frac{x}{x - 1}$

Review
1) $(x + 9)(x - 9)$
2) $(m + 4m)(m - 4m)$
3) $(12 + 11x)(12 - 11x)$
4) $(6b + 2a)(6b - 2a)$

Solutions

Lesson 7-3
Page 137

Exercises
1) -1
2) -1
3) $-\frac{1}{2}$
4) $-(m + 3) = -m - 3$
5) $-(x + 1) = -x - 1$
6) $-\frac{1}{x + 1}$
7) $-\frac{5}{m + 4}$
8) $-\frac{2}{x + y}$
9) -1
10) -2
11) -1
12) -1

13) $-\frac{x + y}{3}$
14) $-\frac{1}{x + 3}$
15) $-\frac{1}{4}$
16) $-\frac{3}{x + 1}$
17) $-\frac{3}{x + 3}$
18) $-(x + y) = -x - y$

Review
1) $(x + 7)(x + 2)$
2) $(x - 2)^2$
3) $(x + 5)(x - 8)$
4) $(x + 4)(x - 9)$

Lesson 7-4
Page 139

Exercises
1) $n = 2$
2) $x = 12$
3) $x = 9$
4) $x = \frac{3}{2}$
5) $y = 8$
6) $n = 3$
7) $n = -3$
8) $x = 8$
9) $x = 3$
10) $x = 4$
11) $x = 11$
12) $x = 7$
13) $x = 20$
14) $x = 3$
15) $x = 10$
16) $x = -10$
17) $x = \frac{4}{3}$
18) $x = 27$

Review
1) $(x + 3)(x - 2)$
2) $(3y + 7)(3y - 7)$
3) $4(x + 5)(x - 3)$
4) $2(2x + 1)(x - 2)$

Lesson 7-5
Page 141

Exercises
1) 2
2) $\frac{1}{x^2}$
3) $6x^2$
4) $\frac{x - 1}{2}$
5) $\frac{2}{3(x - 2)}$
6) $\frac{2a}{x - 2}$
7) $\frac{x}{4}$
8) $2(x + 1)$
9) $\frac{1}{3(x + 2)}$
10) $\frac{2}{3}$

11) $\frac{5}{2(c + 1)}$
12) $\frac{x + 3}{2}$
13) $\frac{x - 1}{2}$
14) $\frac{x}{5(x - 3)}$
15) $9x$
16) $4x(x - 1)$
17) $\frac{3x}{x - 3}$
18) $\frac{x + 5}{5}$

Review
1) $\frac{1}{3y}$
2) $\frac{2a}{x}$
3) $3a^2 b^3$
4) $-5xy^2$

Lesson 7-6
Page 143

Exercises
1) $\frac{5}{4}$
2) 4
3) $\frac{y^3}{x^2}$
4) $\frac{3}{2xy}$
5) $6y$
6) $\frac{9(n - 1)}{4}$
7) $\frac{6(x + 1)}{5}$
8) $\frac{4}{x + 3}$
9) 5

10) $5(x + y)$
11) $\frac{3}{x - 1}$
12) $\frac{x + y}{15}$
13) $\frac{1}{2}$
14) $\frac{x}{2(x - 2)}$
15) $\frac{3}{2}$
16) $\frac{2x - 1}{x(x + 1)}$
17) $\frac{2x(x + y)}{y}$
18) $\frac{xy}{x - y}$

Review
1) $x = 2$
2) $x = 4$
3) $x = \frac{2}{3}$
4) $x = 4$

Lesson 7-7
Page 145

Exercises
1) $\frac{11}{3x}$
2) $\frac{5x - 2y}{9}$
3) $\frac{2}{x}$
4) $\frac{7x}{x + 1}$
5) $\frac{4x + 5}{11}$
6) $\frac{1}{5}$
7) $\frac{x + y}{xy}$
8) $\frac{-2x + 4}{7x} = \frac{-2(x - 2)}{7x}$
9) 2
10) $\frac{5x + 7}{2}$
11) $\frac{2x + 1}{x + 3}$
12) $\frac{1}{x + 2}$
13) $\frac{6x - 10}{5} = \frac{2(3x - 5)}{5}$

14) 1
15) $\frac{1}{x - 2}$
16) $x - 1$

Review
1) $\frac{n - 2}{2}$
2) $\frac{1}{2}$
3) $\frac{m - 4}{3}$
4) $\frac{x - 2}{x + 2}$

Lesson 7-8
Page 147

Exercises
1) $\frac{11x}{4}$
2) $\frac{3x}{2}$
3) $\frac{13}{4x}$
4) $\frac{31}{4x}$
5) $\frac{8y + 1}{6xy}$
6) $\frac{7x}{24}$
7) $\frac{23}{12x}$

8) $\frac{3x}{4}$
9) $\frac{17x}{10}$
10) $\frac{8x - 5}{6}$
11) $\frac{16x + 5}{12}$
12) $\frac{-x + 11}{10}$
13) $\frac{29x^2}{12}$
14) $\frac{8n + 5m}{m^2 n^2}$

15) $\frac{10x + 5}{21} = \frac{5(2x + 1)}{21}$
16) $\frac{3(2x + 3)}{x^2 - 9}$

Review
1) $x = 10$
2) $x = 2$
3) $x = 7$
4) $x = 12$

Solutions

Lesson 7-9
Page 149

Exercises

1) $\dfrac{2x^2 + 5x}{(x+3)(x+2)}$

2) $\dfrac{-x^2 + x - 8}{(x+3)(x-2)}$

3) $\dfrac{9x + 6}{(x+4)(x-1)}$

4) $\dfrac{x^2 - x + 15}{(x-3)(x+4)}$

5) $\dfrac{3x + 14}{x^2 - 9}$

6) $\dfrac{-5x + 14}{x^2 - 16}$

7) $\dfrac{17}{2(x-3)}$

8) $\dfrac{3x - 23}{x^2 - 36}$

9) $\dfrac{2x^2 + 6x}{(x+4)(x+2)}$

10) $\dfrac{-2x - 20}{(x+1)(x-2)}$

11) $\dfrac{x + 8}{x^2 - 9}$

12) $\dfrac{21x + 14}{(x-1)(2x+3)}$

13) $\dfrac{33}{4(x+1)}$

14) $\dfrac{29}{6(x-2)}$

Review

1) $\dfrac{2}{3}$

2) $\dfrac{5a^2}{4b^2}$

3) $\dfrac{x+y}{x-3}$

4) $3(x+1)$

Lesson 7-10
Page 151

Exercises

1) $x = 6$
2) $x = 2$
3) $x = 8$
4) $y = 8$
5) $x = 48$
6) $x = 12$
7) $x = 12$
8) $x = 20$
9) $x = 9$
10) $x = -18$
11) $x = 21$
12) $x = -27$
13) $x = -15$
14) $x = 7$
15) $x = 120$
16) $x = 3$

Review

1) $\dfrac{1}{6}$

2) $6rs$

3) $\dfrac{2}{y+3}$

4) $\dfrac{9(x-1)}{4}$

Chapter 7 Review
Page 152

1) $7x^2 y$
2) $3x + 1$
3) -1
4) $-\dfrac{1}{4}$
5) $x - y$
6) $-(x+4) = -x - 4$
7) $x = 33$
8) $4(x - y)$
9) $\dfrac{m-1}{3}$
10) $\dfrac{4}{x+4}$
11) $n = 10$
12) $-\dfrac{11}{6x}$
13) $\dfrac{2x + y}{8}$
14) $\dfrac{7x - 16}{12}$
15) $\dfrac{x + 7y}{12}$
16) $x = 4$
17) $x = 2$

Lesson 8-1
Page 155

Exercises

1) 20
2) $3\sqrt{5}$
3) $3\sqrt{11}$
4) $10\sqrt{3}$
5) $5x$
6) $2x\sqrt{3}$
7) $3xy\sqrt{5}$
8) $xy\sqrt{y}$
9) $7x\sqrt{x}$
10) $4xy\sqrt{y}$
11) $3xy\sqrt{xy}$
12) $5xyz\sqrt{2}$
13) $4yz\sqrt{yz}$
14) $15x^2 y\sqrt{2y}$
15) $6x^2 y^3\sqrt{2}$
16) $15mn\sqrt{2p}$
17) $x^2 y^2\sqrt{xy}$
18) $3x^4\sqrt{2}$

Review

1) $\dfrac{3+m}{2m}$

2) $\dfrac{19}{12x}$

3) $\dfrac{2b-3}{2ab}$

4) $\dfrac{5}{4a}$

Lesson 8-2
Page 157

Exercises

1) $x = \pm 9$
2) $x = \pm 10$
3) $n = \pm 4$
4) $x = \pm 4$
5) $x = \pm 4$
6) $x = \pm 3\sqrt{5}$
7) $x = 4$
8) $x = 5$
9) $x = 36$
10) $x = 25$
11) $x = 2$
12) $x = 12$
13) $x = 11$
14) $x = 6$
15) $x = 20$
16) $x = 7$
17) $x = 11$
18) $x = 61$

Review

1) $\dfrac{21m - 4n}{9mn}$

2) $\dfrac{8x + 13}{15}$

3) 3

4) $\dfrac{7}{5x}$

Lesson 8-3
Page 159

Exercises

1) $12\sqrt{2}$
2) $5\sqrt{3}$
3) $4\sqrt{3}$
4) $-2\sqrt{7}$
5) $8\sqrt{3}$
6) $4\sqrt{2}$
7) $-6\sqrt{3}$
8) $10\sqrt{3} + \sqrt{5}$
9) $5\sqrt{5} - 4\sqrt{7}$
10) $-3\sqrt{2}$
11) $9\sqrt{3}$
12) 0
13) $\sqrt{2}$
14) $11\sqrt{2}$
15) 0
16) $3x\sqrt{3}$
17) $4\sqrt{x}$
18) $\sqrt{5}$

Review

1) $\dfrac{8x + 5}{x(x+1)}$

2) $\dfrac{x - 7}{(x+1)(x-1)}$

Solutions

Lesson 8-4
Page 161

Exercises

1) 8
2) $2\sqrt{7}$
3) $3\sqrt{7}$
4) $30\sqrt{3}$
5) 36
6) $70\sqrt{6}$
7) 4n
8) $4\sqrt{10}$
9) $3\sqrt{3} - 9$
10) $4\sqrt{3} - 6\sqrt{2}$
11) -1
12) $2 - 2\sqrt{5}$
13) 15
14) $6 - \sqrt{6}$
15) $6\sqrt{2} - 8\sqrt{3}$
16) 1
17) $4 - 4\sqrt{2}$
18) $7 - 4\sqrt{3}$

Review

1) x = 42
2) x = 30
3) n = 10
4) n = 42

Lesson 8-5
Page 163

Exercises

1) 5
2) $2\sqrt{2}$
3) $3\sqrt{3}$
4) 8
5) $8\sqrt{2}$
6) 16
7) x
8) 9a
9) 8
10) $\sqrt{2} + 2$
11) 8a
12) $\frac{3\sqrt{2}}{2}$
13) 25
14) $\sqrt{3}$
15) 9xy
16) 2x
17) $x\sqrt{3}$
18) $3\sqrt{5}$

Review

1) 50
2) $5\sqrt{5}$
3) $2x\sqrt{3}$
4) $5x^2\sqrt{3}$

Lesson 8-6
Page 165

Exercises

1) $\frac{\sqrt{15}}{5}$
2) $\frac{\sqrt{2}}{2}$
3) $\frac{\sqrt{14}}{4}$
4) $6\sqrt{2}$
5) $3\sqrt{3}$
6) $5\sqrt{5}$
7) 9
8) $\frac{x\sqrt{2}}{2}$
9) $3\sqrt{2}$
10) $\sqrt{7}$
11) $\sqrt{3}$
12) $\frac{\sqrt{15}}{3}$
13) $6\sqrt{2}$
14) $2\sqrt{3}$
15) $x\sqrt{y}$
16) 5
17) $\frac{\sqrt{2}}{3}$
18) $\frac{3\sqrt{2}}{2}$

Review

1) n = ±11
2) x = ±10
3) x = ±4
4) x = ±5

Lesson 8-7
Page 167

Exercises

1) $\frac{\sqrt{3} - 1}{2}$
2) $10 - 5\sqrt{3}$
3) $\frac{\sqrt{6} + 6}{-5}$
4) $\frac{2\sqrt{5} - \sqrt{30}}{-2}$
5) $10\sqrt{6} - 20$
6) $4\sqrt{15} + 12$
7) $8 + 2\sqrt{3}$
8) $\frac{4\sqrt{2} + \sqrt{6}}{13}$
9) $\frac{5\sqrt{2} - 15}{-7}$
10) $6 + 2\sqrt{2}$
11) $\frac{7 + \sqrt{7}}{3}$
12) $\frac{12\sqrt{2} + 6\sqrt{10}}{-1} = -12\sqrt{2} - 6\sqrt{10}$
13) $2 - \sqrt{3}$
14) $\frac{1 - 2\sqrt{7}}{-9}$
15) $\frac{\sqrt{5} - 1}{4}$

Review

1) $9\sqrt{2}$
2) $4\sqrt{3}$
3) $\sqrt{3a}$
4) $3\sqrt{3}$

Lesson 8-8
Page 169

Exercises

1) $c = 2\sqrt{5}$
2) $b = 2\sqrt{13}$
3) $c = 5\sqrt{2}$
4) $b = 8\sqrt{2}$
5) $c = 2\sqrt{10}$
6) $b = 2\sqrt{15}$
7) c = 15
8) $c = 2\sqrt{13}$

Review

1) $\sqrt{3}$
2) a
3) $\sqrt{2}$
4) $\sqrt{5}$

Lesson 8-9
Page 171

Exercises

1) 2.8
2) 1.4
3) 3.2
4) 9
5) 5
6) 15.2
7) 9.8
8) 13
9) 11.7
10) 5.8
11) 17.7
12) 5.7
13) 10
14) 5

Review

1) $\frac{\sqrt{3}}{2}$
2) $\frac{\sqrt{10}}{4}$
3) 9
4) $\sqrt{3}$

Solutions

Lesson 8-10
Page 173

Exercises

1) $6, 3\frac{1}{2}$ 10) $2, -3\frac{1}{2}$
2) $4\frac{1}{2}, 1\frac{1}{2}$ 11) $4, 6$
3) $1, \frac{1}{2}$ 12) $1\frac{1}{2}, 5\frac{1}{2}$
4) $0, -\frac{1}{2}$ 13) $-2, -5\frac{1}{2}$
5) $3\frac{1}{2}, 0$ 14) $-7, 8$
6) $2, -2\frac{1}{2}$
7) $6, 2\frac{1}{2}$
8) $\frac{1}{2}, -\frac{1}{2}$
9) $1, 1$

Review

1) $c = 5$
2) $c = 15$
3) $b = 8$
4) $a = 15$

Chapter 8 Review
Page 174-5

1) 30 14) 4
2) $xy\sqrt{5}$ 15) $x = 12$
3) $10x\sqrt{x}$ 16) $x = 7$
4) $x - 5$ 17) 15
5) $x = 11$ or $x = -11$ 18) 9
6) $x = 4$ or $x = -4$ 19) 4.5
7) $8\sqrt{3}$ 20) $(-2, -5\frac{1}{2})$
8) $15\sqrt{3}$ 21) $(-6, -3\frac{1}{2})$
9) $-7\sqrt{3}$
10) $12\sqrt{2}$
11) 180
12) $\frac{x\sqrt{x}}{2y}$
13) $\frac{\sqrt{6}}{4}$

Lesson 9-1
Page 177

Exercises

1) $x = 2$ or $x = 4$
2) $x = -6$ or $x = 1$
3) $x = 11$ or $x = 2$
4) $x = -3$ or $x = 2$
5) $x = 0$ or $x = 5$
6) $x = 0$ or $x = -4$
7) $x = 7$ or $x = -7$
8) $x = 8$ or $x = -3$
9) $x = 3$ or $x = -4$
10) $x = -6$ or $x = 2$
11) $x = 4$ or $x = -3$
12) $x = 2$ or $x = 6$
13) $x = 6$ or $x = -6$
14) $x = 3$ or $x = -9$
15) $x = 7$ or $x = -4$
16) $x = 2$ or $x = 9$
17) $x = 3$ or $x = -1$
18) $x = 3$ or $x = -1$

Review

1) $4\sqrt{10}$
2) $2\sqrt{15}$

Lesson 9-2
Page 179

Exercises

1) $x = \pm 9$
2) $x = \pm 8$
3) $x = \pm 3$
4) $x = \pm 2$
5) $x = \pm 9$
6) $x = \pm 6$
7) $x = \pm 2\sqrt{5}$
8) $x = 9$ or $x = -1$
9) $x = 2$ or $x = -1$
10) $x = 10$ or $x = 4$
11) $y = 1$ or $y = -5$
12) $x = \sqrt{6}$ or $x = -\sqrt{6}$
13) $x = 2$ or $x = -2$
14) $x = \sqrt{5}$ or $x = -\sqrt{5}$
15) $x = 2$ or $x = -6$
16) $x = 10$ or $x = -4$

Review

1) $D = 10$
2) $D = 5$

Lesson 9-3
Page 181

Exercises

1) $x = -1$ or $x = -5$
2) $x = 6$ or $x = -3$
3) $x = 7$ or $x = -1$
4) $x = 5$ or $x = -1$
5) $x = 1 + \sqrt{5}$ or $x = 1 - \sqrt{5}$
6) $x = 6$ or $x = -2$
7) $x = -2 + \sqrt{2}$ or $x = -2 - \sqrt{2}$
8) $x = -1 + \sqrt{7}$ or $x = -1 - \sqrt{7}$
9) $x = -2 + \sqrt{3}$ or $x = -2 - \sqrt{3}$
10) $x = -3 + 2\sqrt{2}$ or $x = -3 - 2\sqrt{2}$
11) $x = -1 + \sqrt{2}$ or $x = -1 - \sqrt{2}$
12) $x = 2 + \sqrt{5}$ or $x = 2 - \sqrt{5}$
13) $x = 3 + 3\sqrt{2}$ or $x = 3 - 3\sqrt{2}$
14) $x = 2 + \sqrt{5}$ or $x = 2 - \sqrt{5}$

Review

1) $(5, 5)$
2) $(5, 2)$
3) $(2, 3\frac{1}{2})$
4) $(3, 4\frac{1}{2})$

Lesson 9-4
Page 183

Exercises

1) $x = 2$ or $x = \frac{1}{2}$
2) $x = -\frac{1}{2}$ or $x = -1$
3) $x = -4$ or $x = 2$
4) $x = 1$ or $x = 2$
5) $x = -1$ or $x = 3$
6) $x = -\frac{2}{5}$ or $x = -1$
7) $x = \frac{-1 + \sqrt{57}}{4}$ or $x = \frac{-1 - \sqrt{57}}{4}$
8) $x = 7$ or $x = -2$
9) $x = \frac{-3 + \sqrt{17}}{4}$ or $x = \frac{-3 - \sqrt{17}}{4}$
10) $x = \frac{-3 + \sqrt{5}}{2}$ or $x = \frac{-3 - \sqrt{5}}{2}$
11) $x = 2$ or $x = -\frac{1}{2}$
12) $x = \frac{1 + \sqrt{5}}{2}$ or $x = \frac{1 - \sqrt{5}}{2}$
13) $x = -\frac{3}{2}$ or $x = -1$
14) $x = -5$ or $x = 1$

Review

1) $y = 0$ or $y = 3$
2) $x = 0$
3) $x = -6$
4) $x = 1$ or $x = -1$

Solutions

Lesson 9-5
Page 185

1)

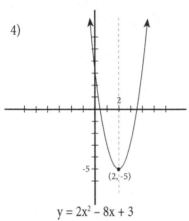

$y = x^2 - 6x + 8$

2)

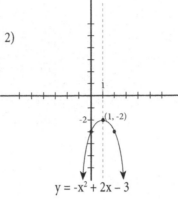

$y = -x^2 + 2x - 3$

3)

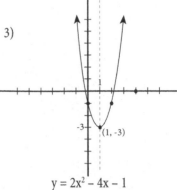

$y = 2x^2 - 4x - 1$

4)

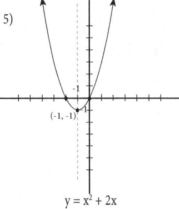

$y = 2x^2 - 8x + 3$

5)

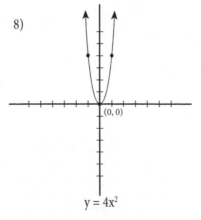

$y = x^2 + 2x$

6)

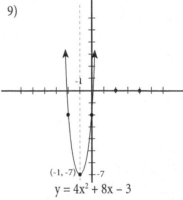

$y = -2x^2 + 4x + 1$

7)

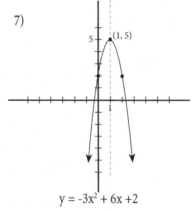

$y = -3x^2 + 6x + 2$

8)

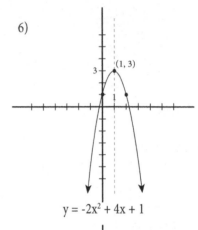

$y = 4x^2$

9)

$y = 4x^2 + 8x - 3$

10)

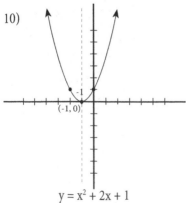

$y = x^2 + 2x + 1$

Review

1) $x = \pm 5$

2) $x = \pm 2\sqrt{2}$

3) $y = \pm 3\sqrt{3}$

4) $n = \pm\sqrt{10}$

Solutions

Lesson 9-6
Page 187

Exercises

1)	2	9)	2
2)	1	10)	0
3)	0	11)	0
4)	2	12)	2
5)	2	13)	2
6)	2	14)	1
7)	2	15)	2
8)	2	16)	1

Review

1) $9(2x + 3y)(2x - 3y)$
2) $5x(2x + 5)$
3) $(x + 3)(x + 4)$
2) $8(3x + 2)(3x - 2)$

Chapter 9 Review
Page 188

1) $x = 0$ or $x = -3$
2) $x = 0$ or $x = -6$
3) $x = \frac{6}{7}$ or $x = -\frac{6}{7}$
4) $x = 5$ or $x = -2$
5) $x = 7$ or $x = 1$
6) $x = -16$ or $x = 5$
7) $y = 4$ or $y = -4$
8) $x = 4$ or $x = -4$
9) $y = \sqrt{15}$ or $y = -\sqrt{15}$
10) $y = 8$ or $y = -8$
11) $x = 2$ or $x = -6$
12) $x = 1 + \sqrt{5}$ or $x = 1 - \sqrt{5}$
13) $x = -1 + \sqrt{7}$ or $x = -1 - \sqrt{7}$
14) $x = 1\frac{1}{2}$ or $x = -3$
15) $x = 3 + \sqrt{7}$ or $x = 3 - \sqrt{7}$
16) 0
17) 2

18)

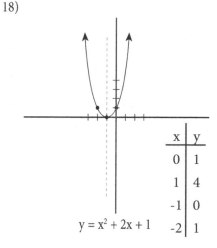

x	y
0	1
1	4
-1	0
-2	1

$y = x^2 + 2x + 1$

Lesson 10-1
Page 191

Exercises

1) $2x - 7 = 12$
2) $3x + 2 = 30$
3) $2x + 5 = 14$
4) $4x - 6 = 10$
5) $4x - 5 = 12$
6) $\frac{1}{3}x - 4 = 2x + 8$
7) $2(x + 2) = 10$
8) $5x - 3 = 17$
9) $2x - 6 = 15$
10) $3x - 2 = 2x + 7$

11) $x + 4 = 7 + -12$
12) $\frac{x}{5} = 25$
13) $4x + 9 = 25$
14) $5x - 16 = x + 5$
15) $4x + 3 = 2x - 2$
16) $2(2y + 7) = 10$
17) $4(x + 9) = x - 8$
18) $\frac{x}{3} + 6 = 14$
19) $\frac{5x}{8} = 12$
20) $4x - 7 = 2x + 12$

Review

1) $x = 5$ or $x = 3$
2) $x = 9$ or $x = -9$
3) $x = 4$ or $x = 1$
4) $x = 5$ or $x = 0$

Lesson 10-2
Page 193

Exercises

1) 8
2) 62
3) 5
4) 3
5) 45
6) 32, 64
7) John $18, Julie $54
8) 555 7th graders,
   340 8th graders
9) Stan 22 candy bars,
   Stan 71 candy bars
10) 62, 63
11) 32, 33, 34
12) 51, 53, 55

Review

1) 22.5
2) 4.8%
3) 60
4) 90

Lesson 10-3
Page 197

Exercises

1) Joe = 7 km/hr,
   Ralph = 8 km/hr
2) 6 hours
3) Fast truck 97.5 km,
   Slow truck 82.5 km
4) 500 km/hr, 560 km/hr
5) 4 hours
6) 36 km
7) 6 hours
8) 8 km

Review

1) $x = \frac{3}{2}$ or $x = -3$
2) $x = 1.5$ or $x = 1$

Solutions

Lesson 10-4
Page 199

Exercises

1) 700 liters of 50 cent cleanser, 300 liters of 80 cent cleanser
2) 20 kg of vanilla, 40 kg of chocolate
3) 60 liters of 24% cream soup, 30 liters of 18% cream soup
4) 20 liters of 30% solution, 40 liters of 60% solution

Review

1) $\frac{3y}{5}$
2) $\frac{20x^2}{y^2}$
3) $\frac{3}{y+3}$
4) $\frac{3(x+1)}{5}$

Lesson 10-5
Page 201

Exercises

1) 2 hours
2) 12 minutes
3) 12 days
4) 6 hours
5) 12 hours
6) Dan works 6 days, Dave works 12 days
7) 3 hours
8) $7\frac{1}{5}$ hours

Review

1) 11
2) 7

Lesson 10-6
Page 203

Exercises

1) Mother 24, Son 4
2) Zach 24, Moe 32
3) Chuck 12, Dave 24
4) Larry 15, Laura 40
5) Lance 17, Sam 24
6) Carly 30, Vica 60
7) Rhoda 6, Sophie 24
8) Ralph 8, Eric 32
9) Son 10, Father 50
10) Olivia 12, Jake 36

Review

1) $x = \frac{1}{2}$ or $x = -4$
2) $x = \frac{7 + \sqrt{33}}{4}$ or $x = \frac{7 - \sqrt{33}}{4}$

Lesson 10-7
Page 205

Exercises

1) 2 dimes, 8 quarters
2) 36 dimes, 24 quarters
3) 5 nickels, 35 dimes
4) 8 nickels, 24 dimes
5) 24 dimes, 16 quarters
6) 12 nickels, 7 dimes, 21 quarters
7) 6 nickels, 24 dimes
8) 24 nickels, 8 dimes, 15 quarters

Review

1) (5, 2)
2) (2, 1)

Lesson 10-8
Page 207

Exercises

1) $4,000 at 6%, $8,000 at 5%
2) $1,400 at 6%, $600 at 9%
3) $4,500 at 6%, $6,000 at 5%
4) $60,000 at 6%, $140,000 at 7%
5) $7,000 at 5%, $9,000 at 6%
6) $8,000 at 6%, $9,000 at 7%

Review

1) $-12x^3 y^3$
2) $-3x^5 y^9$
3) $r^4 s^5 t^5$
4) $20x^6$

Chapter 10 Review
Page 208-9

1) 15
2) 16 and 64
3) 8
4) 56 and 32
5) 10 hours
6) 120 km/hr   180 km/hr
7) 6 hours
8) 180 km
9) 36 kg or $1.80 coffee, 24 kg of $2.80 coffee
10) $18\frac{3}{4}$ minutes
11) 15 hours
12) 16 dimes, 24 nickels
13) 24 dimes, 16 nickels
14) Sophia 10, Maria 15
15) Sonya 54, Harry 18
16) $5,500 at 6%, $3,500 at 7%

Solutions

Chpt. Quiz 1
Page 210-11

1) -284
2) -95
3) -3
4) -168
5) -312
6) -23
7) $-1\frac{1}{12}$
8) $1\frac{1}{8}$
9) 2
10) -6.384
11) 81
12) 7^8
13) 5^9
14) 11
15) 66
16) 2
17) identity of x
18) $2^3, 5^2, 3^3$
19) no - 7 is paired with two y-values
20) 2×3^3
21) $2^3 \times 5^2$
22) 25
23) 120
24) 1.449×10^{10}
25) 1.9×10^{-7}
26) 20
27) 360 miles
28) 42
29) 75%
30) 250 in the school

Chpt. Quiz 2
Page 212

1) x = -23
2) x = 5.4
3) n = -22
4) x = -4
5) x = 54
6) x = -45
7) x = 27
8) x = 3.4
9) x = -4
10) x = 4
11) $x = 6\frac{2}{5}$
12) $x = 2\frac{2}{5}$
13) x = -7
14) x = 0
15) x = -12
16) x = 5
17) x = -5
18) x = 5
19) x = 5 or $x = -7\frac{4}{5}$
20) y = 3 or y = -5
21) 5x – 9y
22) $-3x^2 – 3y – x^2y$
23) x = -14
24) x = 12
25) x = 105

Chpt. Quiz 3
Page 213

1)

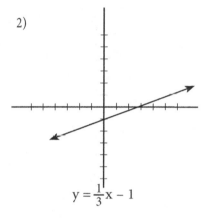

$y = 3x + 1$

2)

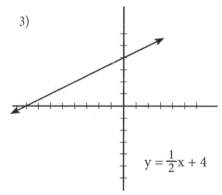

$y = \frac{1}{3}x - 1$

3)

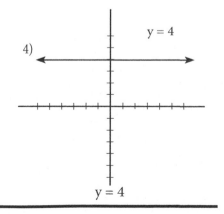

$y = \frac{1}{2}x + 4$

4)

$y = 4$

$y = 4$

Solutions

Chpt. Quiz 3 (cont.)
Page 213

5)
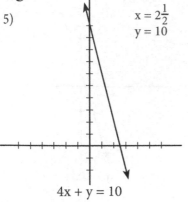

$x = 2\frac{1}{2}$
$y = 10$

$4x + y = 10$

6)
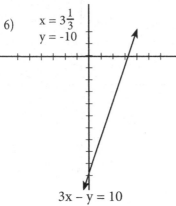

$x = 3\frac{1}{3}$
$y = -10$

$3x - y = 10$

7)
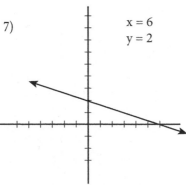

$x = 6$
$y = 2$

$2x + 6y = 12$

8) $-\frac{1}{2}$

9) $\frac{2}{3}$

10) 1

11)
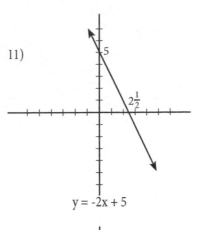

$y = -2x + 5$

12)
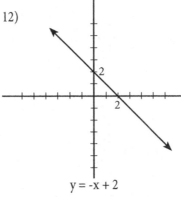

$y = -x + 2$

13)
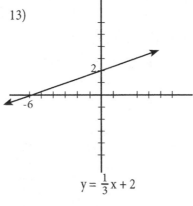

$y = \frac{1}{3}x + 2$

14)
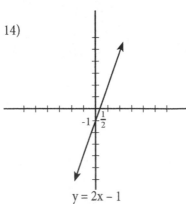

$y = 2x - 1$

15)
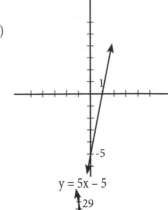

$y = 5x - 5$

16)
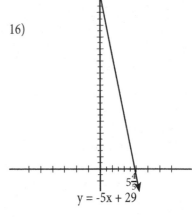

$y = -5x + 29$

17) $y - 2 = 5(x - 3)$

18) $y + 4 = -\frac{3}{4}(x + 3)$

Solutions

Chpt. Quiz 4
Page 214

1)

2)

3)

4) $x \le 2$

5) $x \le -4$

6) $x \ge 10$

7) $x \le -5$

8) $x \ge 5$

9) $x > -3$

10) $x \ge 8$

11) $x > -18$

12) $x > 8$

13) $m < 2$

14) $x \ge 19$

15) $x \ge 3$

16) $-4 < x < 2$

17) $x < 2$ or $x > 7$

18) $-14 < x < 4$

19) $-3 \le x \le 2$

Chpt. Quiz 5
Page 215

1)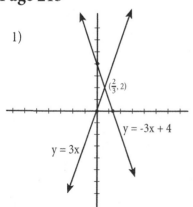

$(\frac{2}{3}, 2)$

$y = -3x + 4$

$y = 3x$

2)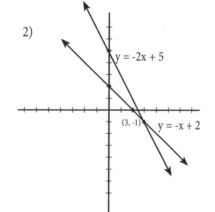

$y = -2x + 5$

$(3, -1)$

$y = -x + 2$

3)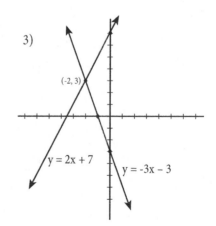

$(-2, 3)$

$y = 2x + 7$

$y = -3x - 3$

Solutions

Chpt. Quiz 5 (cont.)
Page 215

4) $(4, 2)$

5) $(7, 5)$

6) $(12, 4)$

7) $(1, -1)$

8) $(4, 1)$

$y > 2x + 5$
$y < -x$

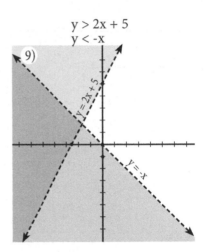
9)

$y > x$
$y > -2x + 4$

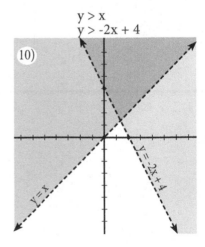
10)

Chpt. Quiz 6
Page 216-17

1) $2y^2 - 6y + 8$

2) $4x^2 + 6x - 2$

3) $3x^2 - 9xy + 8y^2$

4) $x^2 - xy + 6$

5) $27x^6$

6) $-24x^3 y^6$

7) $\frac{6xy}{z}$

8) $8x^3 y^3 - 4x^3 y^2 + 6x^4 y^3$

9) $-66x^4 + 18x^3 + 30x^2$

10) $x^2 + 13x + 42$

11) $40x^2 + 31x + 6$

12) $5x^2 + 3x - 6$

13) $-3xy - x + 2y$

14) $3x + 5$

15) $3x - 8$

16) $7x^2 (4 + 3x^2)$

17) $x (15xy^2 + 25y + 1)$

18) $(x + 5) (x - 3)$

19) $(x - 6) (x + 9)$

20) $(5m + 7) (5m - 7)$

21) $(9y^3 + 5y) (9y^3 - 5y)$

22) $2 (2x + 1) (x - 2)$

23) $10x (x + 2) (x - 2)$

24) $(3x + 1) (x - 2)$

25) $(3x + 1) (x + 2)$

Chpt. Quiz 7
Page 218

1) $\frac{6x^2}{y}$

2) $\frac{3}{xz^2}$

3) $\frac{3}{x + 1}$

4) $\frac{x - 5}{x - 4}$

5) $-(x + 1) = -x - 1$

6) $-\frac{1}{x + 7}$

7) $x = 7$

8) $m = -7$

9) $\frac{2}{5}$

10) $2(x + 2)$

11) $\frac{1}{3}$

12) $\frac{3(x + 5)}{2}$

13) $\frac{11}{2x}$

14) $\frac{4b - 23}{2b + 12}$

15) $\frac{7x}{6}$

16) $\frac{7x}{24}$

17) $\frac{7x^2 + 3x}{(x - 3)(x + 1)}$

18) $\frac{2x - 5}{2x + 5}$

19) $x = 30$

20) $x = 21$

Solutions

Chpt. Quiz 8
Page 219-20

1) $9x$

2) $2x\sqrt[4]{2}$

3) x^3

4) $5xyz\sqrt{3}$

5) $2xy^2z\sqrt{z}$

6) $x = \pm 6$

7) $x = 4$

8) $x = 7$

9) $x = 6$

10) $11\sqrt{5}$

11) $8\sqrt{2}$

12) $4\sqrt{3}$

13) $2\sqrt{7}$

14) 90

15) 72

16) 3

17) $\sqrt{5a}$

18) $4\sqrt{6}$

19) $\dfrac{\sqrt{15}}{3}$

20) $\dfrac{5\sqrt{5}}{2}$

21) $\dfrac{\sqrt{26}}{2}$

22) $\dfrac{12 - 3\sqrt{5}}{11}$

23) $10\sqrt{6} - 20$

24) $c = 65$

25) 10

26) $(0, 0)$

Chpt. Quiz 9
Page 221

1) $x = -5$ or $x = -3$

2) $x = 0$ or $x = -3$

3) $x = 0$ or $x = -8$

4) $x = -5$ or $x = 3$

5) $x = \pm 12$

6) $x = 4\sqrt{2}$ or $x = -4\sqrt{2}$

7) $x = 3$ or $x = -3$

8) $x = 2$ or $x = -1$

9) $x = 7$ or $x = -3$

10) $x = 4$ or $x = 2$

11) $x = -3$ or $x = -1$

12) $x = 3$ or $x = 2$

13) $x = \dfrac{1 + \sqrt{21}}{2}$ or $x = \dfrac{1 - \sqrt{21}}{2}$

14) $x = -1$ or $x = 5$

15) $x = 1 + \sqrt{3}$ or $x = 1 - \sqrt{3}$

16)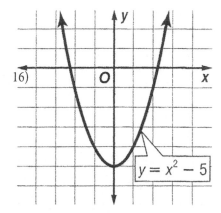

17)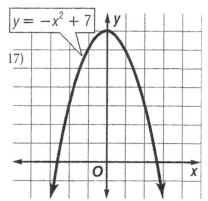

18)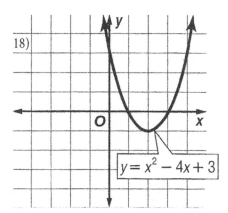

19) 2

20) 0

Solutions

Chpt. Quiz 10
Page 222-23

1) $3x + 12 = 50$
2) $2x + 20 = 48$
3) $5x + 15 = 7x - 3$
4) $6(x - 8) = 15$
5) 15
6) John = 15, Mike = 3
7) 125 and 126
8) 10, 60
9) 4 hours
10) 5 hours
11) $2\frac{2}{3}$ hours
12) 25 lbs. of mixed nuts, 5 lbs. of mixed fruit
13) 12 lbs. peanuts, 8 lbs. of cashews
14) 120 liters of water
15) $3\frac{3}{5}$ hrs. = 3 hrs. 36 min.
16) 24 hrs.
17) 8 hrs.
18) John is 54, Tom is 18
19) Jose is 46, Maria is 26
20) Julie is 40, Kim is 15
21) 3 dimes, 7 quarters
22) 8 nickels, 40 dimes
23) $1200 at 6% interest

Final Review
Page 224

1) -10
2) -20
3) 60
4) 8
5) -3
6) $-\frac{5}{12}$
7) $-\frac{7}{10}$
8) $1\frac{1}{8}$
9) -11
10) -1.4
11) -5.54
12) -18.1
13) 5
14) 343
15) 625
16) 4
17) 7
18) 69
19) 7
20) 2

Final Review
Page 225

21) $2 \times 3^2 \times 7$
22) 10
23) 120
24) 3.21×10^8
25) 2.71×10^{-5}
26) $x = 4$
27) 144
28) 320
29) 80%

Final Review
Page 226

30) $n = -18$
31) $x = 0$
32) $x = -100$
33) $x = 12$
34) $n = 7$
35) $n = -1$
36) $x = 3$
37) $x = 6$
38) $x = 8$
39) $m = 14$ or $m = -2$
40) $x = 4$ or $x = -4$
41) $x = -14$
42) $n = 4$
43) $x = 4$
44) $m = 6$

Solutions

Final Review
Page 227

45)

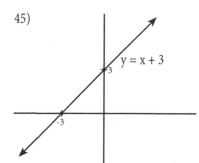

$y = x + 3$

46)

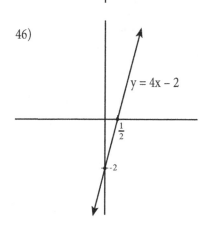

$y = 4x - 2$

47) x-intercept = $1\frac{1}{2}$, y-intercept = 3

48) $-\frac{1}{2}$

49) 1

50) $m = -\frac{3}{2}$

51) $y = -\frac{1}{2}x + 6$

52) $y = 2x + 11$

53) $y = \frac{1}{2}x + 4$

Final Review
Page 228

54) $y - 2 = 3(x - 1)$

55) $-2x + 3y = 12$

56) ←————————•————→
              3

57) ←——∘———•——∘——→
        -2   3

58) ←•———∘————→
      -2   2

59) ←————•———•————→
       -3   3

60) ←—∘————————∘—→
      -4   4

61) $y < -8$

62) $n \geq -5$

63) $x > 20$

64) $-4 \leq x \leq 2$

65) $x < -13$ or $x > 3$

66) $x < -6$ or $x > 6$

67) $-5 < x < 5$

Final Review
Page 229

68)

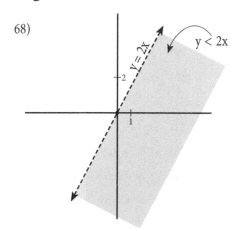

$y < 2x$

$y = 2x$

69)

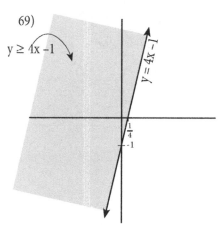

$y \geq 4x - 1$

$y = 4x - 1$

70) (7, 14) $x = 7, y = 14$

71) (5, 1) $x = 5, y = 1$

72) (2, -6) $x = 2, y = -6$

73) (4, -2) $x = 4, y = -2$

74)

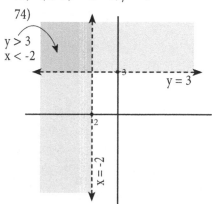

$y > 3$
$x < -2$

$y = 3$

$x = -2$

75) $6x^2y + 4xy^2 + y^3 - 1$

76) $6xy - 3x + 8y$

77) $2x^2 - 4xy - 2y^2$

Solutions

78) $4x^2 + 3xy - 6y^2 - 9$

79) $-30x^4y^6$

80) $20x^5y^6$

81) $-5x^6y^4$

82) $7xy^3z^5$

83) $-12x^3 + 9x^2 - 12x$

84) $6x^4y - 9x^3y^2 - 3x^2y^4$

85) $x^2 + 5x + 6$

86) $10x^2 - 11x - 6$

87) $x - 3x^2y + 2$

88) $2rs + 3s^2 - 6r^2$

89) $3xy(2x - 3)$

90) $x(x - 2)(x - 2) = x(x - 2)^2$

91) $(x + 4)(x + 12)$

92) $(8x + 5y)(8x - 5y)$

93) $5(a + b)(a - b)$

94) $25(x + 4)(x - 2)$

95) $\frac{3}{7}$

96) $\frac{n - 3}{n}$

97) $\frac{x + 3}{x - 3}$

98) $\frac{m - 10}{m + 6}$

99) -1

100) $\frac{-(y + 5)(y - 5)}{7} = \frac{-(y^2 - 25)}{7}$

101) $x = -4$

102) $\frac{5x^2}{4y}$

103) $\frac{x + 2}{2}$

104) $\frac{1}{xy}$

105) $\frac{x - 3}{x + 3}$

106) $\frac{1}{y}$

107) $\frac{-2x + 4}{7x}$

108) $\frac{5}{2x}$

109) $\frac{4y + 1}{6xy}$

110) $\frac{8x + 13}{15}$

111) $\frac{7x}{6}$

112) $\frac{x - 3}{6}$

113) $\frac{12x - 1}{8x^2}$

114) $x = 2$

Solutions

Final Review
Page 233

115) $x = 4\frac{4}{7}$

116) $x = 3$

117) $3\sqrt[3]{7}$

118) $2x\sqrt[4]{7}$

119) $5x\sqrt[3]{3x}$

120) $\frac{x\sqrt[4]{3}}{4}$

121) $\frac{\sqrt[4]{35}}{7}$

122) $\frac{\sqrt[4]{14}}{4}$

123) $\frac{9\sqrt[3]{2}}{4}$

124) $13\sqrt[3]{5}$

125) $4\sqrt[4]{2}$

126) $3\sqrt{2}$

127) 4

128) $6x^2$

Final Review
Page 234

129) $x = 6$ or $x = -6$

130) $x = 5$ or $x = -5$

131) $x = 3\sqrt[3]{5}$ or $x = -3\sqrt[3]{5}$

132) $\frac{\sqrt[4]{3} - 1}{2}$

133) $-\frac{\sqrt[4]{6} + 6}{5}$

134) $c = 3\sqrt[3]{5}$

135) $b = 5\sqrt[3]{3}$

136) $D = 10$

137) $M = (5, 5)$

Final Review
Page 235

138) $x = 6$ or $x = -6$

139) $x = 3$ or $x = -3$

140) $x = -7$ or $x = 1$

141) $x = 8$ or $x = 1$

142) $x = 8$ or $x = 5$

143) $x = -1$ or $x = -2$

144) $x = -5$ or $x = 11$

145) $x = 1 + \sqrt[4]{6}$ or $x = 1 - \sqrt[4]{6}$

146) $x = 2 + \sqrt[4]{14}$ or $x = 2 - \sqrt[4]{14}$

147) $x = \frac{1}{2}$ or $x = -5$

148) $x = 3 + \sqrt[4]{7}$ or $x = 3 - \sqrt[4]{7}$

Solutions

Final Review
Page 236

149)

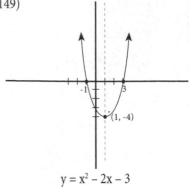

$$y = x^2 - 2x - 3$$

150)

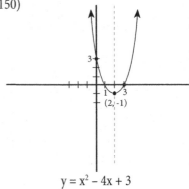

$$y = x^2 - 4x + 3$$

151) 7

152) Vica 40 km, Rick 44 km

153) 1,400 liters of 50 cent cleanser,
     600 liters of 80 cent cleanser

154) $3\frac{1}{3}$ hours

155) $1,200 @ 6%

156) Stan 12, Stuart 18

157) 36 nickels, 12 dimes, 22 quarters

Made in the USA
Monee, IL
22 January 2021